PNSO FIELD GUIDE TO THE ANCIENT WORLD

AGE
OF
ANCIENT SEA
MONSTERS

AGE
OF
ANCIENT SEA MONSTERS

ZHAO Chuang

YANG Yang

BROWN BOOKS
PUBLISHING GROUP

PNSO Field Guide to the Ancient World: The Age of Ancient Sea Monsters

by: Yang Yang, illustrated by Zhao Chuang

Brown Books Publishing Group
Dallas / New York
www.BrownBooks.com
(972) 381-0009

A New Era in Publishing®

Publisher's Cataloging-In-Publication Data

Names: Yang, Yang, 1982- author. | Zhao, Chuang, 1985- illustrator. | Chen, Mo (Translator), translator. | PNSO (Organization), production company.
Title: Age of ancient sea monsters / ZHAO Chuang [illustrator], YANG Yang [author] ; [translator, CHEN Mo].
Description: Dallas ; New York : Brown Books Publishing Group, [2021] | Series: PNSO field guide to the ancient world ; [3] | Translated from the Chinese, published in 2015. | Includes bibliographical references and index.
Identifiers: ISBN 9781612545301
Subjects: LCSH: Marine animals, Fossil. | Evolution (Biology) | LCGFT: Illustrated works.
Classification: LCC QE766 .Y36 2021 | DDC 560.457--dc23

ISBN 978-1-61254-530-1
LCCN 2021908156

Printed in China
10 9 8 7 6 5 4 3 2 1

For more information or to contact the author, please go to www.BrownBooks.com.

Dedicated to

Every living being that enriched the Earth

Contents

9 / The Chaotic Triassic Period

77 / The Silent Jurassic Period

113 / The Volatile Cretaceous Period

FOREWORD

Dr. Mark A. Norell's
Introduction to the Works
by ZHAO Chuang and YANG Yang

Mark A. Norell is a renowed international paleontologist,
Chair and Macaulay Curator for the Division of Paleontology of the AMNH,
and science consultant for PNSO.

I am a paleontologist at one of the world's great museums. I get to spend my days surrounded by dinosaur bones. Whether it is in Mongolia excavating, in China studying, in New York analyzing data, or anywhere on the planet writing, teaching, or lecturing, dinosaurs are not only my interest, but my livelihood.

Most scientists, even the most brilliant ones, work in very closed societies. A system which, no matter how hard they try, is still unapproachable to average people. Maybe it's due to the complexities of mathematics, difficulties in understanding molecular biochemistry, or reconciling complex theory with actual data. No matter what, this behavior fosters boredom and disengagement. Personality comes in as well, and most scientists lack the communication skills necessary to make their efforts interesting and approachable. People are left being intimidated by science. But dinosaurs are special—people of all ages love them. So dinosaurs foster a great opportunity to teach science to everyone by tapping into something everyone is interested in.

That's why YANG Yang and ZHAO Chuang are so important. Both are extraordinarily talented, very smart, but neither are scientists. Instead they use art and words as a medium to introduce dinosaur science to everyone from small children to grandparents—and even to scientists working in other fields!

ZHAO Chuang's paintings, sculptures, drawings, and films are state-of-the-art representations of how these fantastic animals looked and behaved. They are drawn from the latest discoveries and his close collaboration with leading paleontologists. YANG Yang's writing is more than mere description. Instead she weaves stories through the narrative or makes the descriptions engaging and humorous. The subjects are so approachable that her stories can be read to small children, and young readers can discover these animals and explore science on their own. Through our fascination with dinosaurs, important concepts of geology, biology, and evolution are learned in a fun way. ZHAO Chuang and YANG Yang are the world's best, and it is an honor to work with them.

AUTHOR'S PREFACE
Swim in the Oceans with Reptiles

The vast ocean gave birth to countless animals, which constantly appeared in haste and disappeared in sorrow. Over billions of years, some of these animals must have impressed the ocean. Among them, the strange reptiles deserve mention.

Ocean vertebrates spent more than one hundred million years moving from water to land. After colonizing land, some of them unexpectedly returned to the sea. The first movers were reptiles. It was not an easy journey. Some species died before they could adapt to the new environment. Others failed at the brutal competition for survival. Cruel acts and bloodsheds were normal, something to be expected. But could you guess what their struggles eventually led to? They defeated all that was in their path and became the unchallenged hegemons of the ocean.

Age of Ancient Sea Monsters tells the story of how different groups started to move from land to the ocean and adapted to the new environment. We started telling this epic by explaining their need for survival and ended with our appreciation for their courage. It took more than mere courage for them to struggle against and take full control of their fate.

We have depicted an ocean world that seemed alien and hope our readers can be curious and learn about courage. We hope to guide you to these strange creatures and see how they conquered the ocean.

Their unyielding spirit is something that we could learn and pass down to future generations. And, to that end, we invite parents to read this book with their children.

YANG Yang
Beijing, July 2019

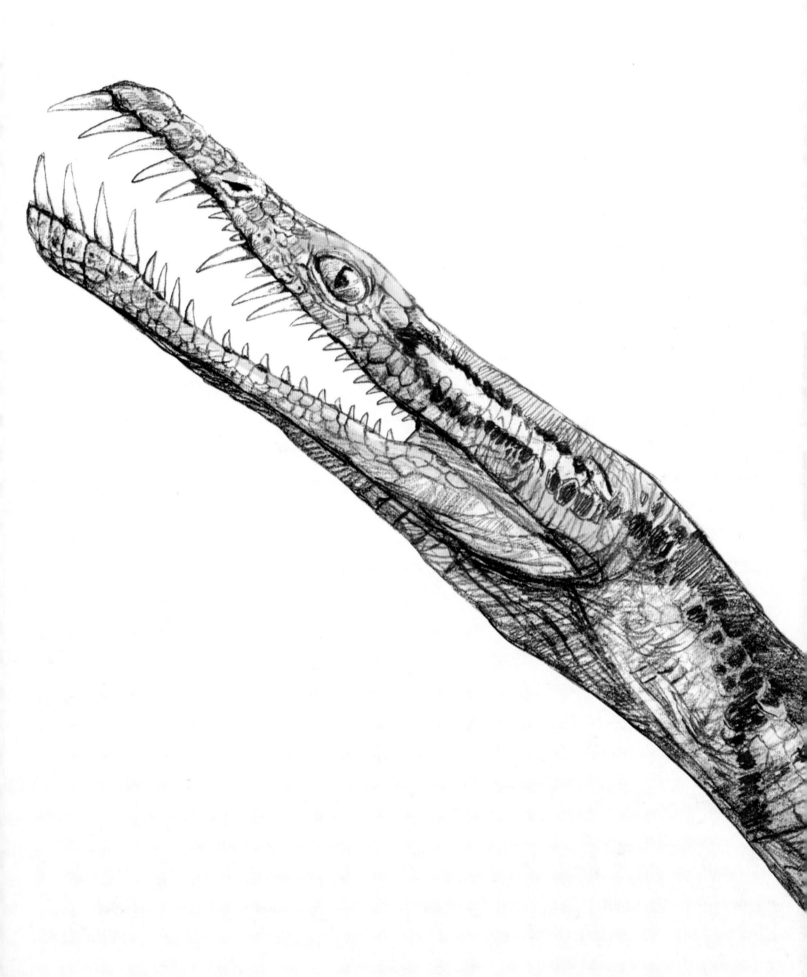

Ambuslocetus

Ichthyostega

Anomalocaris

Hundreds of million years ago

0 — In the year 1602 of our time, the Dutch physicist Cornelius Drebbel created the world's first navigable submarine, allowing modern *Homo sapiens* to return to the deep ocean.

Livyatan — Marine mammals, such as whales, evolved to adapt to a marine lifestyle, occupying the top of the marine food chain.

— Certain mammals moved from the land to the sea.

— Most marine reptiles left the world, and the ocean was once again dominated by fish.

1

2

— Some reptiles returned to the ocean. They had explosive growth in a short period and quickly became new rulers.

Tylosaurus

3

— The earliest four-legged fish, *Ichthyostega*, appeared. This marked the beginning of the vertebrates moving to the land.

4 — Fish appeared. The ten-meter-long *Dunkleosteus* became the new hegemon of the ocean.

Dunkleosteus

5

— The Cambrian explosion brought invertebrates, which ruled the ocean. *Anomalocaris* was an archetype. The oldest vertebrate, *Haikouichthys*, appeared around the same time.

6

35 — The primitive single-celled organisms became the ocean's first inhabitants. Cyanobacteria (blue-green algae), the first single-celled photosynthetic organism, appeared.

Geological Timetable of the Sea Monsters Depicted in This Book

Reference: International Chronostratigraphy Chart 2014
Source: International Union of Geological Sciences (IUGS)
Illustrated by: PNSO

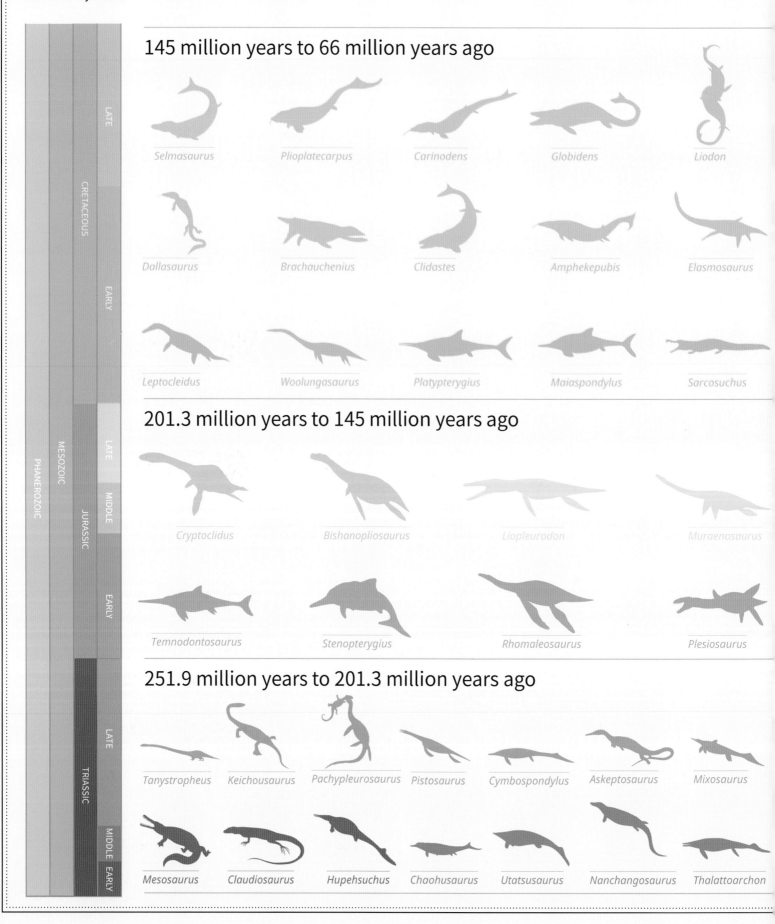

145 million years to 66 million years ago

Selmasaurus   Plioplatecarpus   Carinodens   Globidens   Liodon

Dallasaurus   Brachauchenius   Clidastes   Amphekepubis   Elasmosaurus

Leptocleidus   Woolungasaurus   Platypterygius   Maiaspondylus   Sarcosuchus

201.3 million years to 145 million years ago

Cryptoclidus   Bishanopliosaurus   Liopleurodon   Muraenosaurus

Temnodontosaurus   Stenopterygius   Rhomaleosaurus   Plesiosaurus

251.9 million years to 201.3 million years ago

Tanystropheus   Keichousaurus   Pachypleurosaurus   Pistosaurus   Cymbospondylus   Askeptosaurus   Mixosaurus

Mesosaurus   Claudiosaurus   Hupehsuchus   Chaohusaurus   Utatsusaurus   Nanchangosaurus   Thalattoarchon

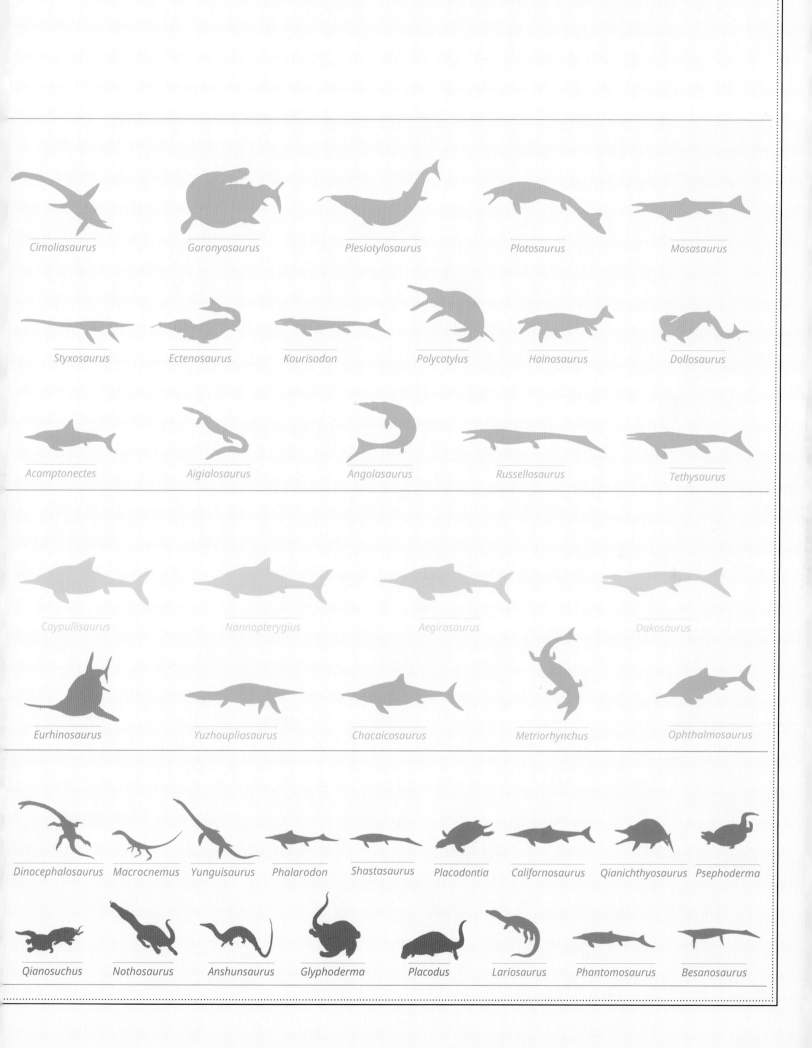

Cimoliasaurus

Goronyosaurus

Plesiotylosaurus

Plotosaurus

Mosasaurus

Styxosaurus

Ectenosaurus

Kourisodon

Polycotylus

Hainosaurus

Dollosaurus

Acamptonectes

Aigialosaurus

Angolasaurus

Russellosaurus

Tethysaurus

Caypullisaurus

Nannopterygius

Aegirosaurus

Dakosaurus

Eurhinosaurus

Yuzhoupliosaurus

Chacaicosaurus

Metriorhynchus

Ophthalmosaurus

Dinocephalosaurus

Macrocnemus

Yunguisaurus

Phalarodon

Shastasaurus

Placodontia

Californosaurus

Qianichthyosaurus

Psephoderma

Qianosuchus

Nothosaurus

Anshunsaurus

Glyphoderma

Placodus

Lariosaurus

Phantomosaurus

Besanosaurus

Origins of the Sea Monster Fossils Referred To in This Book

Illustrated by: PNSO

| ASIA | CHINA | **Present-Day China** | | | |
| | JAPAN | Keichousaurus | Hupehsuchus, Dinocephalosaurus | Chaohusaurus, Macrocnemus | Nanchangosaurus, Yunguisaurus |

NORTH AMERICA	AMERICA	**Present-Day America**	Thalattoarchon	Cymbospondylus	Phalarodon
	CANADA	Dallasaurus	Brachauchenius	Clidastes	Elasmosaurus
	MEXICO	Selmasaurus	Plioplatecarpus	Plesiotylosaurus	Cimoliasaurus

| SOUTH AMERICA | ARGENTINA | **Present-Day Argentina** | Chacaicosaurus | | |

Present-Day Germany

Pistosaurus Phantomosaurus

Placodontia Stenopterygius

Aegirosaurus Dakosaurus

Nothosaurus Platypterygius

Present-Day Italy

Besanosaurus Tanystropheus

Askeptosaurus Psephoderma

Aigialosaurus

Present-Day England

Leptocleidus

Acamptonectes

Liodon

EUROPE: GERMANY, ITALY, ENGLAND, NORWAY, FRANCE, RUSSIA, NETHERLANDS, SWITZERLAND

AUSTRALIA | AUSTRALIA | **Present-Day Australia** | Woolungasaurus

Present-Day South Africa

Moschorhinus

Present-Day Madagascar

Claudiosaurus

Present-Day Angola

Angolasaurus

AFRICA: SOUTH AFRICA, ANGOLA, MOROCCO, NIGER, MADAGASCAR, NIGERIA

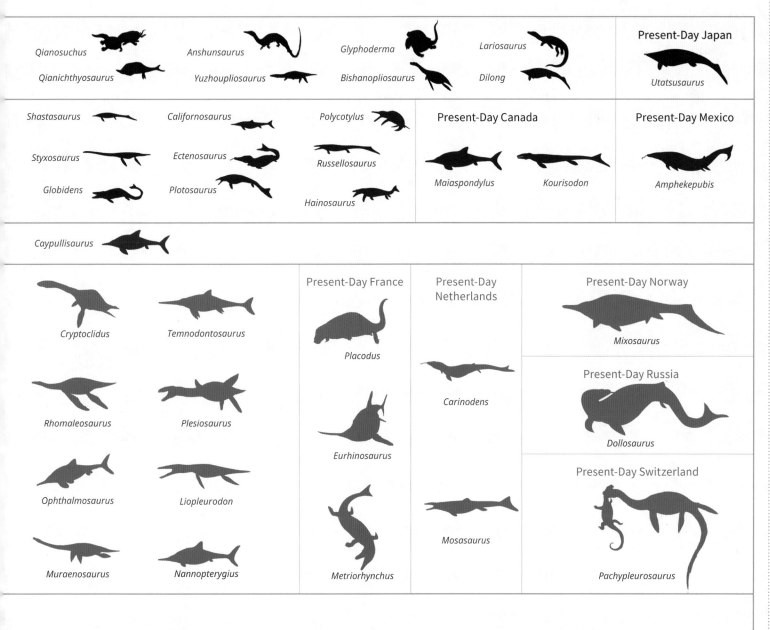

Qianosuchus *Anshunsaurus* *Glyphoderma* *Lariosaurus* **Present-Day Japan**

Qianichthyosaurus *Yuzhoupliosaurus* *Bishanopliosaurus* *Dilong* *Utatsusaurus*

Shastasaurus *Californosaurus* *Polycotylus* **Present-Day Canada** **Present-Day Mexico**

Styxosaurus *Ectenosaurus* *Russellosaurus*

Globidens *Plotosaurus* *Hainosaurus* *Maiaspondylus* *Kourisodon* *Amphekepubis*

Caypullisaurus

Cryptoclidus *Temnodontosaurus* **Present-Day France** **Present-Day Netherlands** **Present-Day Norway**

Placodus *Mixosaurus*

Rhomaleosaurus *Plesiosaurus* *Carinodens* **Present-Day Russia**

Ophthalmosaurus *Liopleurodon* *Eurhinosaurus* *Dollosaurus*

Muraenosaurus *Nannopterygius* *Mosasaurus* **Present-Day Switzerland**

Metriorhynchus *Pachypleurosaurus*

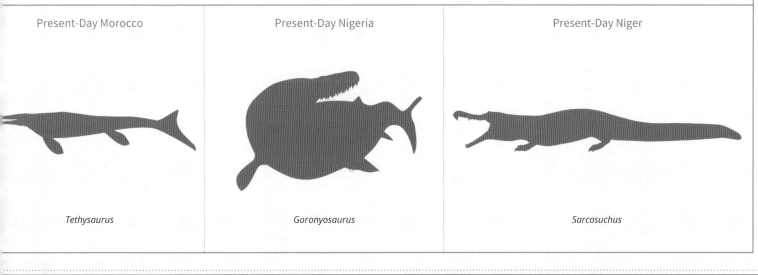

Present-Day Morocco Present-Day Nigeria Present-Day Niger

Tethysaurus *Goronyosaurus* *Sarcosuchus*

The Chaotic Triassic Period

The evolution of vertebrates is unusually complex. Some of them moved from water to land, some wanted to take off from land and fly, and still others got bored with terrestrial life and chose to return to water. In this complicated and long journey, some animals brought revolutionary changes that we now look at in awe. Reptiles were landmark revolutionaries. Since the birth of vertebrates, reptiles were the first animals that could live outside water, and this greatly expanded their living space. However, not all reptiles were satisfied with living on land. Some of them started to move back to water, tracing the footprints of their ancestors.

This sudden move occurred about 251 million years ago, after a global extinction that wiped out most invertebrates in the ocean and more than 70 percent of terrestrial reptiles. The vast ocean was temporarily silent, and the remaining reptiles seized the opportunity to return to the ocean, where food was abundant and competition mild. They tried to revitalize the reptiles in the new environment.

Aquatic vertebrates spent more than one hundred million years to become terrestrial. In contrast, the return to the ocean was easier. The beginning of the Triassic period saw explosive growth of the returned reptiles. In the ocean, as well as in freshwater lakes and rivers, numerous types of reptiles appeared. These included the Protorosauria, Archosauria, Thalattosauria, Ichthyosauria, Placodontia, and Nothosauroidea, all with different forms and habits. They developed rapidly and lived in all parts of the world, dominant in every niche.

However, the chaotic ocean became quiet again at the end of the Triassic period. A few species made it to the Jurassic period, such as the Ichthyosauria order and the Plesiosauria order (the latter evolved from the Nothosauroidea), but the rest of the aquatic reptiles were no more.

Nevertheless, the rapid development of aquatic reptiles in the Triassic period after the Paleozoic extinction remained the most important and most exciting chapter in the global biological recovery.

500 Million Years Ago
Present-Day America

The vast sea ushered in its first living being four billion years ago, but the real flourishing of life came only three-and-a-half billion years later.

The ocean flowed and thundered alone on planet Earth, day after day, year after year. It remembered every bump and trough on the seafloor, every tiny single-celled life, but it took a long time for more guests to arrive.

It was a long, discouraging wait. It seemed that simple lives were all that was possible in the vast ocean. But then, the show began.

No one knew where the guests were coming from, but they suddenly filled the ocean. They soon crowded the ocean, each looking for a seat.

These were the marine invertebrates, small animals armed with hard shells, and they soon got the best seats.

The ocean smiled. The pushing and hustling were fine. It was a long wait, and these guests deserved a long-lasting banquet that never ends.

But who were those huge guys? These giants were not talking to anyone, and they scared away the trilobites next to them.

The ocean heard that these were the *Anomalocaris*, who would later dominate all other animals for a long time. The ocean felt that it needed to have a word with them. This was a party, and everyone should be lively. These silent guys should join the fun.

400 Million Years Ago
Present-Day Morocco

When someone disturbed the darkness, and the sun hurried to illuminate their path, all inhabitants of the ocean knew that the *Dunkleosteus* was coming.

The *Dunkleosteus* arrogantly wiped off the water that was blocking its eyes, swam rapidly, and let the sunlight shine upon the smaller creatures. The small fish and crustaceans were facing an apocalypse. Trying to run away, they picked the shortest path away and stirred up the whole ocean. They had no choice because they saw how terrifying the *Dunkleosteus* was. As long as it was willing, the hunter could effortlessly open its mouth and suck all of them into the stomach in one-fiftieth of a second.

The *Dunkleosteus* got a bit angry, but the three of them did not intend to punish these small, timid guys right away. It was heading towards a big shark hiding in the darkness.

The ocean felt terrifying. The shark must have sensed it and wanted to escape, but it was too late.

Death was quick. The male *Dunkleosteus* opened its mouth and bit hard, and the victim's blood instantly reddened the ocean water. The female *Dunkleosteus* and its child could enjoy the delicious meal without joining the fight.

The era of the marine invertebrates soon ended. The oldest fish, *Haikouichthys*, appeared in the Cambrian period. Since then, the huge family of fish evolved to dominate the ocean with lightning speed, and they remained there for a long period of time. Eventually, some of them got bored with the calm sea and wanted to explore the unfamiliar shore.

The rest is history. Amphibians and reptiles appeared on the earth. They left the ocean, the cradle of life, to explore new horizons.

280 Million Years Ago
Present-Day South Africa

A tree trunk, not yet rotten, lay in the middle of a stream. This was the watchtower for a *Mesosaurus*.

Mesosaurus was a minority among Permian reptiles. About 95 percent of reptiles at the time lived on land, while the *Mesosaurus* belonged to the aquatic 5 percent. Its limbs still resembled reptiles, but webs had grown between its toes. Like paddles, the webbed feet propelled its slender body through the water. Its mouth was covered with sharp comb-like teeth, which could filter small fish in the water.

It climbed the watchtower and struggled to raise its head up. Mocking laughter came to its ears, "What could you see? Nothing more than some tree crowns and some useless path that leads to nowhere!"

The *Mesosaurus* never responded to such meaningless taunts. It knew that what it saw would slowly build up its world. Moreover, it smelled something strange coming from afar—fresh, moist, and a bit salty.

Where did the smell come from? The *Mesosaurus* wanted to know.

255 Million Years Ago
Present-Day Madagascar

A *Claudiosaurus* and its companions were playing in the shallow sea.

If *Mesosaurus* could meet the *Claudiosaurus,* it would have its question answered. The unusual smell came from the ocean, where the *Claudiosaurus* lived. Unfortunately, a major extinction occurred between the end of the Permian period to the Early Triassic period, killing more than 70 percent of Earth's reptiles. The *Mesosaurus* and its family were not spared.

Mesosaurus, who lived in smaller streams, could not have imagined the changes to come in thirty million years. By then, new reptiles moved from the land to the ocean, something that only happened in *Mesosaurus*'s wildest dreams. In its dream, the water would no longer be a gentle stream but would gush and surge as it got carried by the strongest wind, with the little reptile standing on top of the waves like a conqueror.

The dream must have been fulfilled by the *Claudiosaurus* and others who successfully survived the extinction disaster.

Undoubtedly, the *Claudiosaurus*, who lived in the Late Permian period, was fortunate. It spent most of the time resting on land, but when it had the energy, it would go into the ocean to enjoy some brief leisure.

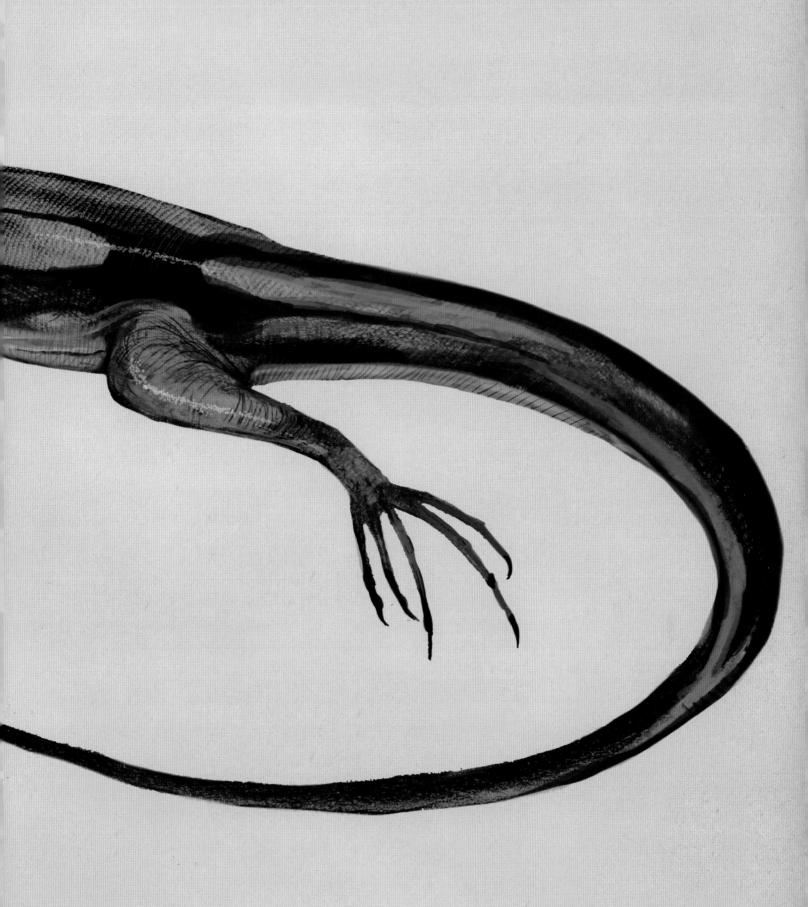

250 Million Years Ago
Present-Day China

It was as if the *Claudiosaurus* opened the door to a new world. Shortly after its appearance, the world entered a new era, the Triassic, and many reptiles started a new life in water.

Most of the earlier marine invertebrates were killed in the extinction, and not many terrestrial reptiles on the land were spared. The surviving reptiles knew that their ancestors spent hundreds of millions of years moving to the land from the ocean, but still, they dreamed of re-entering the ocean, now as quiet as if newly created. They knew that if they were successful, they would again dominate this paradise.

Even with the bright future in its sight, the timid and cautious invertebrate would only stand and watch. Making a change was difficult. Only real fighters could carry it out.

But 250 million years ago in present-day Hubei Province, China, an animal resembling a crocodile appeared in the ocean. It was covered with scaly skin, and it had webbed feet and a sharp, slender mouth.

It looked like a weird-looking submarine as it searched for prey in the ocean.

It was a mysterious *Hupehsuchus*. As of today, people had no idea of how it was related to other marine reptiles, except that it was a close relative of *Nanchangosaurus*.

245 Million Years Ago
Present-Day China

The *Hupehsuchus* was like a hermit traveler who suddenly appeared in a desert. No one knew where it came from or where it was going. It was a free wanderer, alone in the vast world.

The *Chaohusaurus*, however, was different. It was born with a mission 245 million years ago in present-day Anhui Province, China.

The petite *Chaohusaurus*, only seventy centimeters long, unexpectedly moved into the erratic ocean. The ocean was still quiet, and no one knew that its small step will soon change the entire ocean's ecosystem, leading to the birth of the most important group, the Ichthyosauria order.

245 Million Years Ago
Present-Day Japan

The sun ran through the ocean to its bottom, bringing light to the mysterious depths.

If one traced the sunlight, one would find the *Utatsusaurus*, which swam leisurely in the ocean.

As sunlight moved, the *Utatsusaurus* swam to chase it. Its streamlined body and small flippers were better adapted to living in the ocean than the family's predecessor, *Chaohusaurus*. That explained its merry mood in playing with the sunlight.

With the bodies of its members adapting to living in the ocean, the ichthyosaurs took solid steps towards ruling the ocean.

245 Million Years Ago
Present-Day China

A *Nanchangosaurus* looked up at the ocean's surface, feeling good.

As a close relative of *Hupehsuchus*, *Nanchangosaurus* was also a rare sight in the ocean. It looked different from other animals swimming around because it looked half-crocodile, half-ichthyosaur. It had a slender head and a thick tail. Its back was not as smooth as an ichthyosaur but had the crocodile's bony scales. This protection allowed it to swim without worrying about ferocious predators. It did not fret about being different but enjoyed it, knowing that only the brave could afford to be non-conformist.

244 Million Years Ago
Present-Day America

Eight million years after the tragic extinction in the late Paleozoic era, *Thalattoarchon*, an eight-and-a-half-meter-long giant appeared in the ocean. This was a sign of nature's vitality. The giant proved the ichthyosaurs' strength in creating a miracle in the recovery of marine lives.

The hungry *Thalattoarchon* opened its mouth, looking for food eagerly in the ocean. Its teeth were large, with sharp serrated edges. No prey would dream of escape from its attack.

The huge *Thalattoarchon* undoubtedly was the ocean's top predator. If it did not prey on someone, it was only because the latter did not taste nice. This sounds cruel, but that has always been the law of nature, where the strong prey on the weak with no compassion.

240 Million Years Ago
Present-Day China

The Ichthyosauria order was not the only one that colonized the ocean. Another fierce-looking reptile had also begun to explore the new territories. It was the Archosauria order.

Two hundred forty million years ago in present-day Guizhou Province, China, the *Qianosuchus* pushed its body forward by moving its webbed limbs against water. Soon, a prey moved towards it head-on. It opened its mouth, full of sharp teeth, and bit against the belly of the prey. The prey was struggling in panic, to no avail.

The ocean was a fighters' paradise, where predation was easy. The *Qianosuchus* always filled its stomach in a short time, so it could enjoy the warm sunshine on the shore for the rest of the day.

A short adventure secured a long-lasting good life. This was the life for those who made revolutionary changes, not for those who were content with their present.

240 Million Years Ago
Present-Day Germany

In present-day Germany, 240 million years ago, the ocean also had excellent predators, who were as good as the *Qianosuchus*. This time, the protagonist was the *Nothosaurus*.

The family which the *Nothosaurus* belonged to was different from the *Qianosuchus*'s. They were more like modern seals, and they were about three to four meters long. They had long necks, small heads, long tails, and limbs like paddles. They belonged to the Nothosauroidea order.

The *Nothosaurus* was one of the founders of the family. It was petite but fierce, strong enough to be a pioneer.

It always tracked the prey slowly and attacked when the time was ripe. Few could escape its mouth, which was full of teeth as sharp as steel needles.

240 Million Years Ago
Present-Day China

The adventures of the ocean-returning pioneers quickly spread to the shore, and more animals were eager to enter that broader environment. Of course, most of them heard the stories, got excited about the achievements, but automatically ignored the dangers and hardship. Perhaps that was how they, originally timid and cautious, ignited the flame for greater glory.

The Thalattosauria order belonged to the new challengers.

An *Anshunsaurus* was moving carefully in the shallow sea, exploring the unfamiliar world with its slender, lizard-like body. Suddenly, it heard something strange. It stopped, stepping its webbed foot on a log lying on the seafloor sea, looking around cautiously.

The *Anshunsaurus* came from the Thalattosauria order. It had a slender head and neck. Its long tail could push its body forward like a paddle.

240 Million Years Ago
Present-Day China

It was lunchtime. A *Glyphoderma* took a break from its swimming exercise to look around for food. It looked like a present-day tortoise, with its body covered with solid scales. However, it had a long tail, something that the modern-day tortoise lacks.

The two-meter-long *Glyphoderma* was larger than a tortoise, but it was a mediocre swimmer. Still, it had been practicing to adapt to the ocean better.

Look, its practice was effective. It used its wide feet to half-crawl, half-swim. Soon, it found its favorite food. It could then enjoy a good lunch.

The *Glyphoderma* belonged to the Placodontia order, which were exotic marine reptiles. Most of them had a carapace and looked like tortoises, and they had wide, flat bodies and short necks. Some had a plastron in addition to the carapace. Despite their similar appearance, the *Glyphoderma* were not biologically related to tortoises. *Anshunsaurus* belongs to the Thalattosauria order. It had a slender head and neck and a long tail that propelled the body forward like a paddle.

240 Million Years Ago
Present-Day France

If much of the Placodontia order were like tortoises, what about the others? Well, you would get the answer by looking at the *Placodus*.

The weather was excellent: there was no wind, and the ocean was calm. A *Placodus* was taking a leisurely stroll. Its body was a little fat, and its waist was thick, making it look like a swollen lizard. Still, it tried its best to look graceful. It was mediocre at swimming, so it was almost walking in the clear water. Of course, it did not move fast, but it got help from its flat tail and short legs. More importantly, it had the leisure time—time to slow down and enjoy life!

The *Placodus* represented another type of the Placodontia order. They resembled lizards with short and strong limbs.

Soon, the *Placodus* found its favorite prey. Since it was a good idea for a little snack before continuing to walk, it used its prominent front tooth to snatch the hard shell of the shellfish, pulled it out from the ocean floor, and then crushed the shell with the flat teeth behind. It could then enjoy a delicious meal.

237 Million Years Ago
Present-Day China

The *Nothosauroidea* order that emerged in the Early Triassic period quickly adapted to the new environment. By the Middle Triassic period, the family flourished and firmly settled in the warm shallow seas of Guizhou Province, China.

It was already afternoon, but the *Lariosaurus* had not had any food. It was a rare event, for it had been roaming unchallenged in this shallow sea for a long time.

The *Lariosaurus* was from the Nothosauroidea order. Its appearance was peculiar, with the forelimbs already evolved into flippers, but the hind limbs still had five toes each. It had a short neck and small flippers, so it was an average swimmer at best. After hunting and eating, it usually rested on the shore.

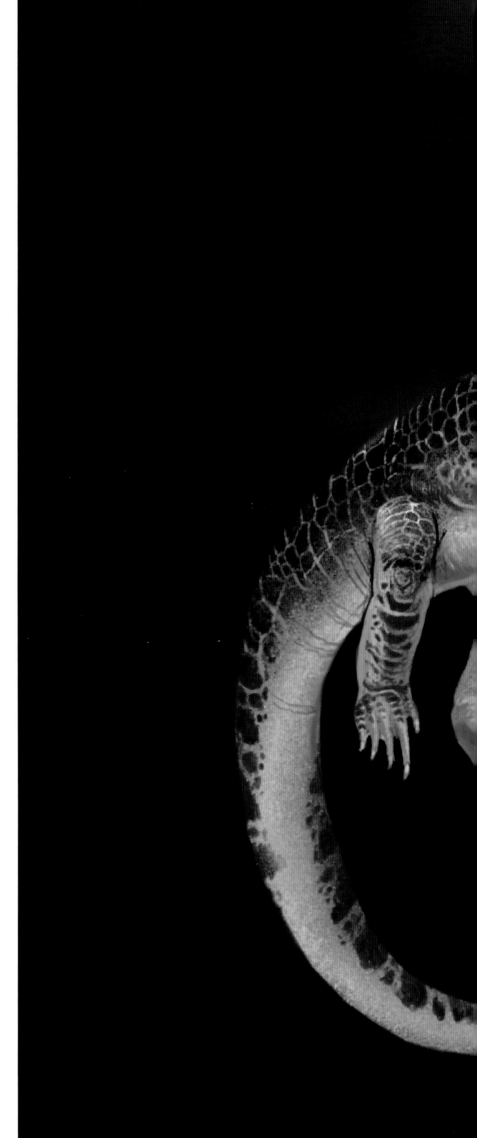

237 Million Years Ago
Present-Day Germany

We have seen members from the Thalattosauria, Nothosauroidea, and Placodontia orders. Compared to them, the Ichthyosauria order was even more ambitious.

Although most of the early ichthyosaurs were petite and lived by preying on small fish and crustaceans, they were determined to conquer the wider oceans. Born in the Early Triassic period, the Ichthyosauria order had its first explosive growth in the Middle Triassic.

The *Phantomosaurus*, a member of the family, appeared during this period. It was so named because its fate was like Erik, the hero in *Phantom of the Opera*. Erik had a face that even his mother was afraid to look at. In order to survive, he hid his face behind a mask. Like Erik, the *Phantomosaurus* also had an ugly head. This was not how its head actually looked when it was alive, however; people who were supposed to keep the fossil safe mishandled and allowed acidic substances to erode the skull fossils, turning it to a scary monster. The *Phantomosaurus* was actually an elegant swimmer.

Two hundred thirty-nine million years ago in present-day Germany, a *Phantomosaurus* was cheerfully moving in the ocean, not expecting that millions of year later, its reputation would be tarnished because of someone else's mistake.

235 Million Years Ago
Present-Day Italy

The ocean was surprisingly quiet and glowing like sapphires. A *Besanosaurus* and its companions were swimming freely, enjoying this wonderful time. Furthermore, it had a lovely baby in its belly, ready to be born. Its eel-like body was a little excited, and it moved the four flippers, making splashing sounds. Those who swam near it could feel its happiness.

This sea was a paradise for marine animals. In addition to *Besanosaurus*, others lived here, too. They were the small sharks, primitive *Osteichthyes* (bony fish), and marine invertebrates. With its narrow head and small tail fins, *Besanosaurus* were still primitive, but they were not small. They could grow to six meters long, with a slender neck and streamlined body. Their large fin-like limbs helped them swim. Although they swam slowly, they used ambushing tactics to make them the top predators in this paradise.

232 Million Years Ago
Present-Day Italy

The ichthyosaurs, like the *Besanosaurus*, stood at the top of the food chain. They had few competitors but maintained peace and were willing to share the ocean with other animals. They, too, would like some friends so that they would not be bored to death!

Among those who shared the living space with them, there was the Protorosauria, a reptile order that had existed since the Permian period. Most of the Protorosauria had evolved into arboreal animals, leaving only a few being aquatic, like the Tanystropheidea family.

The name Tanystropheidea means "long hinge," and it told us that they must have long necks. This *Tanystropheus*, living in present-day Italy 232 million years ago, had a neck that stretched to three meters, longer than the rest of the body.

Its peculiar figure seemed to make its movement in water somewhat troublesome. Most of the time, it walked in the water on its short limbs. It had a long head with staggered teeth, which appeared to be excellent tools for fishing.

The *Tanystropheus* were large enough such that they rarely became food for ichthyosaurs. Rather, the two got on well, and ichthyosaurs would feel less lonely with their company.

230 Million Years Ago
Present-Day China

In the waters of present-day Guizhou Province, China, 230 million years ago, apart from the *Lariosaurus*, the *Keichousaurus* shared this tranquility.

Keichousaurus belonged to the Pachypleurosaur suborder, which were marine reptiles closely related to the Nothosauroidea family. Its bone structure was light and looked like that of a lizard. They had only recently adapted to living in water and could swim fast by using their webbed limbs and slender tails. But they could not swim far into the deep ocean; often, they enjoyed the comfort in the shallow waters near the shore.

One morning 230 million years ago, a *Keichousaurus* found a comfortable rock and gracefully prostrated on it, as usual. The *Keichousaurus* did not have many relatives, and this one particularly enjoyed being alone. In sunny mornings such as this one, or in the warm afternoons, or quiet dusks, one would find this creature by the beautiful shores. It was never in a hurry to hunt because the shallow sea had abundant food. It preferred to spend time listening to and looking at the world in front of it.

At the moment, it happened to hear a wonderful sound, coming from the wind roaring through the cycads on the shore.

230 Million Years Ago
Present-Day Switzerland

In the warm shallow seabed, a cluster of corals gave off a brilliant luster. Attracted by their bright colors, a *Pachypleurosaurus* gently swam to them and stopped to watch.

The corals moved rhythmically with the flow of seawater, like a rainbow hanging in the sky after a rain. The *Pachypleurosaurus* apparently had forgotten about its daily chores. Its eyes were wide open, and its lips formed a smile.

At this moment, a *Simosaurus*, a member of the Nothosauroidea order, was quietly approaching. This was not a graceful swimmer but a fearsome predator nevertheless. It moved slowly towards the opponent, who was far smaller and outmatched, fully confident that it would make a catch.

The *Pachypleurosaurus* was still happily viewing the beautiful scene, completely unaware that soon it was to be devoured by someone else. This was a cruel world.

230 Million Years Ago
Present-Day Germany

The breath of lives was spreading in the ocean. It was like a strong current, like the ice in late winter or soil in early spring. More and more animals traced that current. They looked forward to breaking away from their current constraints to explore and experience new powers and opportunities in the vast sea. By doing so, they forgot about how painful it would be to go through this evolution.

In the early afternoon, a wind rushed to the ocean, stirring the water violently. The wind disturbed the clouds' nap, and the unhappy cloud turned grey in the blink of an eye.

The water was still calm underneath, and the *Pistosaurus* from the *Pistosauridae* family wanted to rest, to gather strength for tonight's hunt. But coincidentally, an unfocused squid was swimming in its direction. *Pistosaurus* was tempted. The squid was larger than its usual target, but it was an opportunity not to be missed. Making up its mind to attack, it instantly mobilized every cell in its body. Hunting meant death for one side but survival for the other, and the *Pistosaurus* always enjoyed surviving. Having prepared itself, it shot out like an arrow.

The frightened squid jumped out of the sea, the *Pistosaurus* followed, and a brutal battle began.

230 Million Years Ago
Present-Day America

The *Cymbospondylus* was the one that made the ichthyosaurs important residents in the ocean.

The appearance of the *Cymbospondylus* was mediocre. Its look gave the impression that it was not particularly suited to living in the ocean. They had small flippers, tiny caudal fins, and no dorsal fins; they were slim, and in trying to swim faster, they always twisted and swayed their bodies like a sea snake, but that only made them look clumsy. They were not small; these giants could grow to six to ten meters, but they could not threaten some of the smaller, one-meter-long nothosaurs.

The description seemed to imply that their existence would be temporary. But that was not the case.

Even with all these downsides, the *Cymbospondylus* succeeded in expanding their living territories to all oceans of the world and becoming one of the most widely distributed ichthyosaurs.

Most of us tend to believe in only what we see, but the fact is sometimes hidden in a corner, where we must spend time digging it out.

A *Cymbospondylus* was happily swimming to its companions. The family was flourishing, and this one had countless friends.

230 Million Years Ago
Present-Day Norway

If you have learned something from *Cymbospondylus*, you will think twice before commenting on this one.

You will think hard before judging which of the things you see is true, which is a distraction.

"It's a wise thing to do!"

Well, I didn't make this condescending remark. The *Mixosaurus*, which lived 230 million years ago in present-day Norway, did.

The *Mixosaurus* was foraging in the ocean, looking relaxed and confident. It was merely one meter long, shorter than some of the water grass, but this short guy was a powerhouse in the vast ocean.

The *Mixosaurus* and the *Cymbospondylus* lived in the same era, and they both lived in the world's every ocean. If you decided that they were weak because their body size was petite, you would have made a grave mistake. In fact, their small bodies were optimally structured. They were shaped in the same way as fish, they had dorsal fins on their backs, and caudal fins took half of the length of their bodies. These were the powerful driving forces which spread the *Mixosaurus* to the world.

230 Million Years Ago
Present-Day Italy

In a short time, the desolated ocean became lively. In the beginning, food was plentiful, and the returning reptiles spent some leisurely time. But soon, they had to figure out some special tricks so that they could take an advantageous position in the increasingly fierce competition.

Askeptosaurus, like the *Anshunsaurus*, belonged to the Thalattosauria order. It successfully expanded its territory to the deep sea, a quieter and safer place compared to the shallow sea. The deep sea is dark, but you need not worry about them. The *Askeptosaurus* had huge eyes and good vision so that they could see through the endless darkness.

This was 230 million years ago in present-day Italy, when an *Askeptosaurus*, who had been enjoying the solitude in the deep sea, raised its head above water to greet the long-lost world.

228 Million Years Ago
Present-Day China

Two hundred twenty-eight million years ago, a strange marine reptile, the *Dinocephalosaurus*, lived in present-day Guizhou Province, China. They figured out a trick that was different from the *Askeptosaurus*. They developed a long neck to deal with prey easily.

This fishing *Dinocephalosaurus* showed how the long neck worked. It first set its eyes on a small fish that was about to swim to its eyes. Without moving its body, it opened its mouth, sucking the small fish in. Its long neck acted like a vacuum cleaner. Despite being inflexible, the neck could suck in the desired food, by expanding and contracting the muscles on its esophagus.

Of course, this long neck also often acted as a good tool for ambushing prey. The long neck allowed its fat body to be kept away from the prey, so the *Dinocephalosaurus* always made full use of it in hunting.

228 Million Years Ago
Present-Day China

The *Dinocephalosaurus* and *Tanystropheus* looked very similar, because both belonged to the *Tanystropheidae* family, and all family members had long necks. We are going to meet another member, the *Macrocnemus*.

Two hundred twenty-eight million years ago in present-day in Guizhou Province, China, a *Macrocnemus* and its companions took leisure in sunbathing on the shore. The sun brightened the patterns on their skin.

They were resting to recover energy, and sunbathing was their way to release stress. In order to win one competition after another, taking a short break was essential.

228 Million Years Ago
Present-Day China

Two hundred twenty-eight million years ago in present-day Guizhou Province, China, a *Yunguisaurus* was flapping its four flippers. Swimming from afar, it moved slowly in the ocean, like an elegant lady. This was an eye-opening trip, and it was curious about everything.

However, some scenes made it feel uncomfortable. Whenever it saw a bloody killing, it closed its eyes and fled quickly.

Yunguisaurus came from the Pistosauroidea clade, and they were related to famous nothosaurs. However, *Yunguisaurus* preferred to be alone than to be in a crowd. It did not hang out with its relatives often, so even when it traveled to the territories of the nothosaurs, it would not tell them in advance.

225 Million Years Ago
Present-Day America

If you like to complain about your hometown, environment, or the people you meet, don't do it. Life could be pretty, but it could also be willfully nasty. We had to adapt to it. The *Phalarodon*, which lived in present-day America 225 million years ago, understood this.

The *Phalarodon* was a member of the Ichthyosauria order. They were small, only one to three meters long, and easily fell prey to others. They were not particularly fat, but neither were they excellent swimmers. A crescent-shaped caudal fin would be ideal to live in water, but their fins were not like that, so the swings of their caudal fins were insufficient to push their body forward.

These features probably tell us that they should not have become the dominant species in the family. On the other hand, they were extremely capable of adapting to the environment, and they used to live in present-day North America, Asia, and Europe. Moreover, they had two types of teeth in their mouths: the small, sharp teeth were ideal for eating fish, while the thick and round ones could easily crush the hard shell of shellfish. This feature allowed them to eat a greater range of food and improve their chances of survival.

Therefore, they have lived leisurely in the ocean for almost twenty million years, longer than many dominant species.

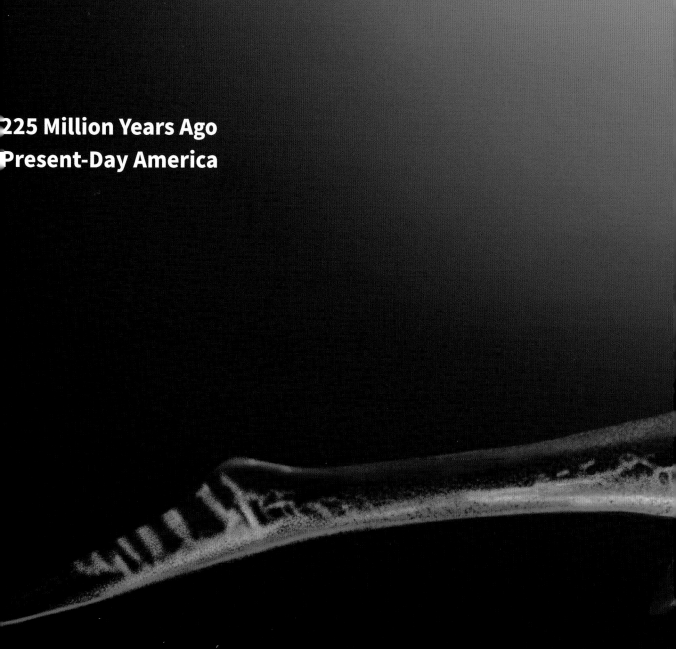

225 Million Years Ago
Present-Day America

Compared with the *Phalarodon*, the *Shastasaurus* had a more advanced body structure and was a better adapter. They lived in present-day America, 225 million years ago, and were the rulers of this part of the world.

They had perfectly streamlined bodies with well-developed flippers, which gave them excellent steering and balance, such that they could move quickly in the water. They could quickly see things around them with a keen vision, and they would attack quickly when they were in danger. They stood at the top of the food chain, such that they not only enjoyed feeding on easy prey but also acted like responsible kings, fighting off intruders to defend their home.

The *Shastasaurus* became the big stars of the Ichthyosauria order in the Triassic period by virtue of their strength. They were one of the most widely distributed species of ichthyosaurs

225 Million Years Ago Present-Day Germany

Throughout the Triassic period, marine reptiles developed explosively, with new species evolving as if a creator ordered them to do so. This magnificent, dramatic global recovery followed the extinction of the Paleozoic and was unexpected in many ways. Even the Placodontia order, whose body structure prevented them from becoming great swimmers, made remarkable progress. In the Late Triassic, one could find Placodontia everywhere: in present-day Europe, North Africa, the Middle East, and China's Guizhou Province.

Two hundred twenty-five million years ago, in present-day Germany, a *Placochelys* was looking for food at the bottom of the water.

With its body covered by a carapace, it looked like a turtle, and a big one—about one meter long. It liked to eat mollusks and crustaceans, and its mouth was a good tool for crushing the hard shell of its prey. Food was becoming scarce these days, but the *Placochelys* was not depressed or angry, knowing that losing its temper could not solve any problems.

220 Million Years Ago
Present-Day America

Although more and more animals were living in the ocean, the ichthyosaurs had become increasingly dominant. The change in their body shape helped them maintain that dominance. They had put in some hard work to adapt to living in the ocean.

Two hundred twenty million years ago in present-day America, a *Californosaurus* was swimming towards a light beam in the ocean.

The beam was from the sun, near the surface of the ocean. The ocean gave a dazzling, sapphire-like glow. The tempted *Californosaurus* twisted its dolphin-like body and rushed into the light. It was swimming and dancing, shattering the sapphire into glittering pieces.

The *Californosaurus* represented an unprecedented change in the ichthyosaurs. Unlike the traditional, stout, lizard-like bodies of the previous ichthyosaurs, the *Californosaurus*'s body was more like a perfectly streamlined fish. It grew dorsal fins, and its caudal fins were larger. All of which heralded the ichthyosaurs' rule of the ocean, the new glory for this ancient family.

220 Million Years Ago
Present-Day China

The ocean is a beautiful place, attracting one group of tourists after another. It dazzles them and makes them speechless. The tourists stop here for a while, take photos, say poetic or romantic words, and leave. As long as they make no attempts at disturbing the lives of the animal residents of the ocean, the latter would probably treat their arrival as a harmless curiosity. But hundreds of millions of years ago, there were no people on the Earth and no other travelers who needed not to worry about food. Therefore, the *Qianichthyosaurus* living in the deep sea had no visitors, and life was a bit lonely.

The *Qianichthyosaurus,* a special member of the ichthyosaurs, was eagerly chasing a small fish. Unlike other members, its body was not beautifully streamlined. It was fat, with its back raised high, and had no dorsal fins. It was only about two meters but not easily captured, thanks to its extremely large eyes.

The eyes of the *Qianichthyosaurus* were larger than all other ichthyosaurs, and their diameters could reach 8.3 cm. This made seeing prey easier and allowed them to see in darkness. Therefore, the *Qianichthyosaurus* could live in the deep dark sea, where there were few competitors, and food was unbelievably abundant.

Now, the *Qianichthyosaurus* was full, and it did not need to eat the small fish it was chasing. It was playing a hunting game because a game would add some fun to life.

210 Million Years Ago
Present-Day Italy

Marine reptiles born at the end of the Early Triassic period experienced explosive growth and spread their members to everywhere in the world. They were confident that their dominance would continue, but their fortune declined in the Late Triassic period. The Ichthyosauria order and the Plesiosauria order, one that evolved from the Nothosauroidea order, lived through the Triassic period and continued to thrive in the Cretaceous period. Apart from them, the rest thinned out and gradually disappeared.

The *Psephoderma* was half-crawling, half-swimming with its wide feet, on its way to eat a barnacle, its favorite. It could grow to nearly two meters long. With a hard carapace covering its back and a rib cage protecting its belly, any enemy would find it a difficult target to attack.

The *Psephoderma* was one of the last members of the Placodontia order, and it witnessed the decline of the family.

The Silent Jurassic Period

In the Early Jurassic period, many aquatic reptiles still lived in the vast oceans as well as in the freshwater rivers and lakes. But the number of species declined, and most of them were from either the ichthyosaur or the plesiosaur families. Compared with the chaotic Triassic period, the Jurassic was almost silent.

The ichthyosaurs were large aquatic reptiles that looked like fish or dolphins. They were not the first reptiles to enter the ocean, but one of the most successful. The Ichthyosauria order had diverse species with varying sizes: some of the smaller ones were two meters long, and the largest could grow to more than twenty meters. They were widely distributed, living in large groups in present-day western Europe, North America, and southern China. They lived in the world for about 150 million years. During this period, they constantly made breakthroughs themselves to adapt to living in the water, eventually climbing to the top of the food chain and becoming the ocean's true hegemons. However, by the end of the Cretaceous period, the glory of the Ichthyosauria order was no more, and their dominant position was replaced by the emerging Plesiosauria order.

The plesiosaurs evolved from the earlier Nothosauroidea order. Its two suborders, the long-necked Plesiosauroidea and the short-neck Pliosauroidea, were dominant species that adapted well to aquatic life. Soon after they appeared, they successfully spread themselves all over the world and replaced the Ichthyosauria order as hegemons. They maintained their superiority until the Cretaceous period, when the *Mosasauridae* family emerged. Despite this, the Plesiosauria order continued to live until the end of the Mesozoic era.

Some other aquatic reptiles also shared the Jurassic ocean with these two groups, but the ichthyosaurs and the plesiosaurs were far more dominant. The constant contest between the two shaped the balance of the entire Jurassic ocean.

198 Million Years Ago
Present-Day England

The Triassic period had given way to the Jurassic. The ocean remained lively, but it had a different group of inhabitants. Almost half of the newcomers were ichthyosaurs, who had their moments of loneliness, embarrassment, doubts, and despair, but they also welcomed comfort, merriment, hope, and beauty. Their number was never as many as that at the second great explosion of the Late Triassic period, but they looked completely different from the primitive look of the earlier members. They were round and looked like present-day dolphins, with a body shape to help them adapt to life in water.

With its long, pointed snout guiding its way, the *Temnodontosaurus* was looking for its favorite food in the ocean. From time to time, small fish were swimming past it, but its big eyes would not bother to give them the slightest attention. It had a strong head and a powerful jaw, both of which allowed it to capture larger prey. Therefore, it never made the lives of the little ones miserable.

189 Million Years Ago
Present-Day Germany

It was game time for the *Stenopterygius*. It swam to look for where the ocean was colored. Could the seawater be colored? Yes. Refraction of sunlight can paint some parts of the sea with colors. Apart from the attentive *Stenopterygius*, few would discover this secret. It did not take long before finding that familiar and colorful scene, which was like a "rainbow" falling into the ocean, and it stopped, getting ready to enjoy its playground.

The *Stenopterygius* adjusted its body and dashed into the "rainbow." Then, in a split second, its round body was jumping out of the ocean, carrying beautiful splashes, to come out under the bright sunlight.

It was finding colored seawater so that it could get closer to the sun. This was the game played by the *Stenopterygius.*

The *Stenopterygius* was representative of the ichthyosaurs during the Jurassic period. Its body was beautifully streamlined, with a dorsal fin and large caudal fins, both of which gave it enough power to propel quickly in the water. When it moved at high speed, the short flipper-like limbs steered its direction. They were about two to four meters long, and they relied on speed to survive in the ocean's fierce competition.

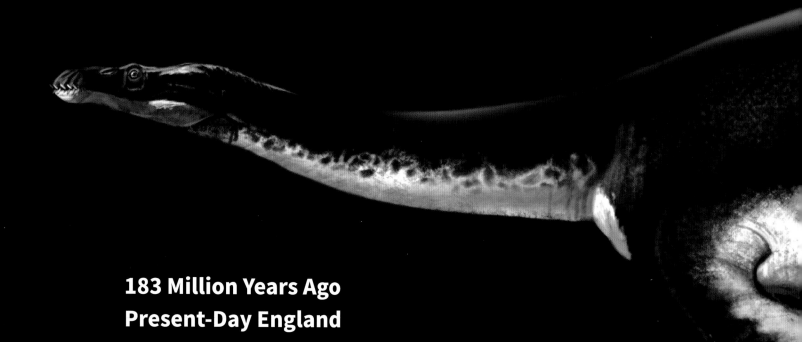

183 Million Years Ago
Present-Day England

After the extinction event at the end of the Triassic period, the plesiosaurs shared the ocean with the surviving Ichthyosauria order. Evolved from the earlier Nothosauroidea order, the plesiosaurs could can be divided into two categories, which differed in their looks: one suborder, the Plesiosauroidea, had a small head and a long neck, living in the shallow sea; the other one, the Pliosauroidea, had large heads, and short, narrow necks, and they lived in the deep sea. We are going to meet the *Rhomaleosaurus*, a member of the Pliosauroidea.

One hundred eighty-three million years ago, in present-day England, a *Rhomaleosaurus* was looking for food in the water.

When a *Rhomaleosaurus* hunted, it held its mouth slightly open. Water would flow into its mouth and then come out of its nostrils. By doing this, it could find the smell of prey in the flowing water. It preferred this rather than using its eyes.

Although the *Rhomaleosaurus* was from the Pliosauroidea, it looked different from those family members that appear later. Its neck was longer and its head smaller, making it look somewhere in between the Plesiosauroidea and the Pliosauroidea.

180 Million Years Ago
Present-Day England

For the whole morning, the gigantic *Plesiosaurus* was out of breath, chasing a small fish. It was a funny look.

This was what happened. The *Plesiosaurus* did not intend to start hunting so early, but a naughty small red fish swam around it when it yawned. Thinking that it was an easy catch, the plesiosaur decided to work for an early breakfast. It opened its U-shaped mouth, trying to swallow the fish in one gulp. However, the little red fish was much more agile, and it slipped away quickly. This was irritating, so it flapped its fours flippers to chase.

The small fish swam fast and dodged the plesiosaur's attacks. The *Plesiosaurus* was also a fast swimmer, but it was big and not as flexible. Sometimes, it seemed that it was about to catch up after a long pursuit, but the little fish took a quick turn and went the other way. If the *Plesiosaurus* could lift its neck, it would quickly gobble the fish, but its long neck was stiff and could not move at all. After a long and disappointing pursuit, it had to let the fish escape.

Like the *Rhomaleosaurus*, the *Plesiosaurus* was also from the Plesiosauria order, but it belonged to the Plesiosauroidea suborder. It had a small head and a long neck. It had no caudal fins and used its flippers to push itself forward while the tail was only used in steering.

180 Million Years Ago
Present-Day France

The vast ocean seemed to hoard unlimited food, but hunting was not as easy as it seemed. Naturally, no one wanted to be prey! Therefore, even the ichthyosaurs, the hegemons of the ocean, had to come up with new tricks to make hunting easier.

A "sharp snout demon" was living in the ocean 180 years ago in present-day France. It was the *Eurhinosaurus*, and it had a fearsome reputation.

The *Eurhinosaurus* had a long, arrow-like maxilla that was almost three-fourths the length of the skull. The maxilla was covered with sharp teeth to hunt.

At the moment, the *Eurhinosaurus* spotted a ray. It used the maxilla to constantly stir up the mud and sand, destroying the ray's home, forcing it to come out of the sand. This was an easy catch with no effort spent on the pursuit.

If a larger predator wanted to prey the *Eurhinosaurus*, it would be difficult. The *Eurhinosaurus* had a streamlined body with large flipper fins and caudal fins. These features made it an excellent swimmer, able to get out of danger quickly.

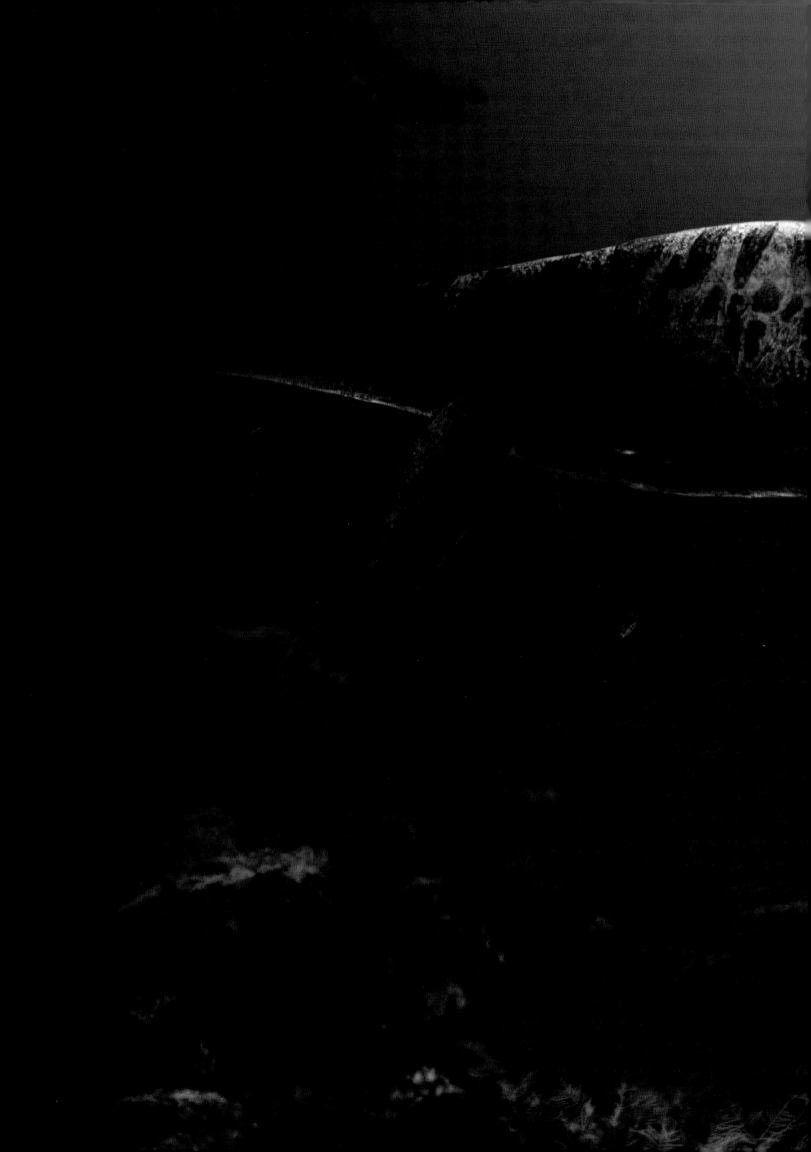

170 Million Years Ago
Present-Day China

The lake was unusually calm, so eerie that it induced horror. The facial expression of the *Yuzhoupliosaurus* was also scary.

At the bottom of the rocky lake, the *Yuzhoupliosaurus* quietly wandered, with its sharp teeth protruding out of its mouth, warning anyone who was getting close. Its body length was four meters, not large compared to other inhabitants in this lake, but it was fierce and aggressive, and being smaller than its opponents was no excuse to avoid a conflict.

The *Yuzhoupliosaurus* was thinking about the big battle. The opponent had agreed to fight. It gave looks to make the whole lake feel tense and terrifying.

In the Jurassic period, the plesiosaurs mostly lived in the ocean of present-day Europe, but the *Yuzhoupliosaurus* was an exception. Compared to the raging sea, it preferred to dominate the calmer fresh water.

170 Million Years Ago
Present-Day Argentina

When not hunting prey, the animals in the ocean were quiet. They swam gracefully in the seawater, appreciated the scenery, looked for lovers, or lazily drifted without doing anything. They enjoyed the good times.

The *Chacaicosaurus* was enjoying a wonderful afternoon alone, as usual. Usually, it swam for some time and then stopped to look around, always finding something new even though it lived in the same region every day.

This afternoon, the new thing that the *Chacaicosaurus* discovered was not scenery, but a beautiful female *Chacaicosaurus*. She must have come from a distant place as she looked tired. The male waited quietly on the side, not wanting to disturb it. The male would like to say hello after the beautiful lady has fully rested.

The *Chacaicosaurus*'s size was medium and had no peculiar features. The big secret was in its mouth. The young *Chacaicosaurus* had tiny sharp teeth, useful for catching fish, but after adulthood, these teeth disappear, and the toothless adult *Chacaicosaurus* preferred to prey on squid.

167 Million Years Ago
Present-Day France

In addition to the ichthyosaur and plesiosaur families, a group of efficient predators was also active in the Jurassic ocean. They were the Thalattosuchia clade, the marine crocodiles. In the Triassic, scale-wearing crocodiles appeared, which were exclusively terrestrial animals. In the Early Jurassic period, much of the land was submerged underwater, and some crocodiles started to live amphibious or marine life, thus becoming marine crocodiles.

Since the *Metriorhynchus* moved into the sea, the original residents were living in constant fear. They stayed at home all day, no longer dared to roam around freely. Even when they had to go out for food, they must make sure that the area was clear, that the *Metriorhynchus* was not nearby. The inhabitants were not timid, but the *Metriorhynchus* was formidable. They no longer wore heavy armor, their limbs had evolved into flippers, and their tails were like fins. They ate all kinds of prey, could swim quickly and swallow the babies of fish, ammonite, or plesiosaurs in no time. Sometimes, they would even jump out of the water to capture small pterosaurs or large *Leedsichthys*.

Right now, the old inhabitants could not deal with this monster, apart from being patient and running away.

165 Million Years Ago
Present-Day England

We always believe that seeing is believing, but sometimes we don't see the whole truth.

Living in the blue sea was a famous star by the name of *Ophthalmosaurus*. Its teardrop-shaped body was beautiful, but that did not explain its fame; other family members were not bad-looking either. It was well known because others thought it mysterious. No one seemed to have seen it hunting. It naturally did not need to eat anything, maybe. Was this true?

Late at night, the tired ichthyosaurs were no longer paying attention to the strange *Ophthalmosaurus*, and all fell asleep. The *Ophthalmosaurus* had not slept yet. It opened its big eyes to look around. Suddenly, it noticed that something was moving, and it dived. Within a few minutes, it returned, as if nothing had happened.

The *Ophthalmosaurus* had a pair of large eyes that allowed it to see prey clearly in the deep dark sea. Thanks to its eyesight, it always chose to prey on a quiet night, when its companions were sleeping soundly. Even in their wildest dreams, they would not think about hunting at night.

165 Million Years Ago
Present-Day England

The plesiosaurs were developing fast, and sometimes even they found the pace of change uncomfortable.

After a nice nap, the *Cryptoclidus* felt refreshed, fully recovered from the fatigue of the past few days. It had traveled from another distant sea to this place, and its original intention was to make a short stay before continuing its journey. However, after arrival, it was attracted by the comfortable life and planned to settle. The sea was calm and beautiful, no one was rushing to get a kill. Everything was slow, and one could even hunt elegantly.

Others would think that the *Cryptoclidus* was a fierce predator. It could eat any prey it wanted to. But no one knew that it was gentle and desired a peaceful life.

The *Cryptoclidus* gently moved its body, trying to make itself as slow and peaceful as the old inhabitants. Still, it failed to hide its large body and sharp teeth. As much as it wanted to be part of its new home, upon seeing it, the schools of fish still scattered in fear.

The *Cryptoclidus* was already an advanced predator, such that it no longer worried about hunger. But compared to the earlier family members, making friends was becoming more difficult. The sullen *Cryptoclidus* returned to its new home, hoping to make some new friends in the future.

164 Million Years Ago
Present-Day China

The evening was the start of a masquerade. Everyone rushed to the makeup artist, the sun, which would give them new looks. The sky arrived the earliest, and the magic sun turned it into a colorful oil painting. Then, the ocean followed. The clear blue sea water was turned cyan. Then, again, there were all the animals. They pushed each other to compete for a good position. A *Bishanopliosaurus*, who had been busy all day, also swam to the surface. It had a remarkable catch, and its stomach was full. This was a good time for it to relax, and it hoped that the sunlight could go through the water to give it a perfect look.

The *Bishanopliosaurus* was as mighty as anyone in the Pliosauroidea suborder. It had long and sharp teeth and could catch fish easily, dominating the rivers and lakes.

160 Million Years Ago
Present-Day England

In the Jurassic period, the plesiosaurs quickly became dominant. Not only did they quickly spread the family into waters of the world, but they also grew rapidly in size. In a short time, many large species were born, especially in the ocean.

The *Liopleurodon* quietly hid in the water, waiting to ambush prey. The *Liopleurodon* was a well-known monster in this area. It was seven meters long and a fierce giant. Even more frightening, it was always hiding in a dark corner, attacking when the prey was unprepared, giving a fatal blow. Residents called it "the dark ambusher."

At this moment, it was carrying out a dark plan one more time. The experienced assassin was once again hiding in the darkness. A *Eustreptospondylus* walking on the shore did not know that. It was leisurely looking for drinking water.

The *Eustreptospondylus* finally chose a good spot, where the land was low and it was easy to drink from the cool river. As it lowered its head to drink, the *Liopleurodon* rushed out of the water, jumping at the prey.

The *Liopleurodon* opened its terrifying mouth and bit into the *Eustreptospondylus.* The victim wanted to resist, but it was too late!

160 Million Years Ago
Present-Day England

The *Muraenosaurus* also lived in present-day England one hundred sixty million years ago. Compared to the *Liopleurodon*, *Muraenosaurus* was gentler. Despite being a good hunter, as good as the *Liopleurodon*, it paid more attention to the beautiful things in life, and by doing so, it became less ferocious and less bloody.

Now, it accidentally found a blue-purple coral, newly appeared on the path that it traveled on every day. Perhaps the coral had always been there, but it never noticed. The corals were not exceptional, but they were as blue as the clear sky, glowing like a purple crystal.

The *Muraenosaurus* was attracted by the elegance. It did not try to touch the coral, but closed its eyes, trying to feel the beauty as if it was afraid of disturbing the coral's peace.

150 Million Years Ago
Present-Day Argentina

Despite being prosperous in the early Jurassic period, the ichthyosaurs had not always been lucky. They did their best to improve all parts that they thought possible, but the Pliosauroidea suborder, a formidable competitor, arrived. The fierce Pliosauroidea had huge bodies and sharp teeth, and in a short time, they destroyed the ocean's delicate balance like a tornado. Since the Middle Jurassic period, the ichthyosaurs showed apparent signs of decline. Although several of them extended their habitat to present-day South America, by the end of the Jurassic period, only a handful of species were active in the ocean.

A *Caypullisaurus* and its companions living in the deep sea were waiting for an opening to hunt. Its body was already well adapted to living in water: its wide and powerful caudal fins and four stout fins could provide enough power for its quick attack, and its big eyes could notice any unusual movement. However, all these did not seem to improve its luck. Too many ferocious animals were living in the ocean. The *Caypullisaurus,* only three meters long, had to hunt in groups to catch even the smaller fish.

148 Million Years Ago
Present-Day England

Apart from present-day South America, Europe in the Late Jurassic period was another refuge for the ichthyosaurs. They tried tirelessly to gain back some advantage against the winning plesiosaurs.

The *Nannopterygius* was an unusual member of the ichthyosaurs. It had very short limbs, which were less than half the size of its head. If you think that it was giving up fights, think again. The three-meter-long creature was too small to win head-to-head fights in a highly competitive environment. To survive, it reduced the size of the limbs to swim faster, such that it could rely on its speed in hunting. In addition, the high speed was its defense against larger predators.

If you could go back to the Late Jurassic ocean and find an animal dashing around, moving its four small flippers, it must be the small *Nannopterygius*.

145 Million Years Ago
Present-Day Germany

The contest between the ichthyosaurs and the plesiosaurs continued. The ancient ichthyosaurs had a strong will and never gave up, but the enemy was too strong to overcome by spiritual strength alone. If you look at the petite *Aegirosaurus*, which was only two meters long, you would not think of it as a fighter.

A nervous *Aegirosaurus* was swimming around in the ocean.

"Can you stop? At least for a little while. My head is spinning by looking at you going in circles," its companion asked.

"No, no, there are dangerous guys everywhere. If I am not careful, I will become food in their stomachs." The other *Aegirosaurus* shuddered.

"But you have sharp teeth to protect yourself," its companion reminded it.

"Against those terrible monsters? My teeth were only useful for scratching their backs," the *Aegirosaurus* said as it swam faster. Speed was probably its only trick against the fierce enemies.

145 Million Years Ago
Present-Day Germany

While the ichthyosaurs and plesiosaurs fought for dominance in the ocean, the marine crocodiles flourished. They quickly adapted to living in the ocean, spread their members globally, and joined the ranks of the top predators.

A pterosaur was hovering in the cloudless sky, flying up and down. Suddenly, a fierce *Dakosaurus* jumped out of the ocean. Even the pterosaurs knew that this was a fearsome predator. The pterosaur thought that the *Dakosaurus* was picking it as a target to attack, so it quickly fluttered its wings and flew up. After a while, it turned back and saw that the *Dakosaurus*'s mouth was holding a small ichthyosaur, which was struggling in vain to dodge the bites of the *Dakosaurus.*

The pterosaur realized that before it came to the sea, the *Dakosaurus* had already started a fierce hunt, and the pterosaur was not the target.

The pterosaur was still trembling in the air, feeling lucky that it had narrowly escaped death. Seeing that the *Dakosaurus* had already torn the ichthyosaur into pieces with a hundred huge teeth, the pterosaur was feeling sad about the victim but happy that it did not have to fight this enemy.

The Volatile Cretaceous Period

After entering the Cretaceous period, the few remaining members of the Ichthyosauria order were no longer ruling the waves. The powerful plesiosaurs replaced them to become the hegemon of the ocean. But they did not celebrate the victory for long. Soon, the fierce *Mosasauridae* family appeared.

In billions of years, the vast ocean ushered in one group of powerful predators after another. They came and went in a hurry, leaving behind their glories and regrets. Many of these are blurred, but the ocean never forgets the top predators that lasted from ninety-three million years ago to sixty-six million years ago. They were the mosasaurs of the *Mosasauridae* family.

The mosasaurs were aquatic reptiles with curly bodies that looked like snakes. They had been around for less than thirty million years, short by the standard of prehistoric animals, but they were the most ferocious residents ever witnessed in the ocean.

When mosasaurs appeared, the Cretaceous ocean was no longer calm. They turned almost all the animals into their own prey, including members of the same family and the unfortunate dinosaurs who inadvertently walked near the water.

The mosasaurs had perfect bodies. In a few million years, the little lizards evolved into terrible monsters. Until now, no one has broken that record. They were undoubtedly the perfect hegemon.

Still, the *Mosasauridae* family became extinct about sixty-six million years ago, not because they somehow failed to adapt to the marine environment, but due to the terrible extinction of the end of the Cretaceous period. All hegemons of the world, the mosasaurs and plesiosaurs in the ocean, dinosaurs on land, and pterosaurs in the sky, disappeared, leaving only their endless legends.

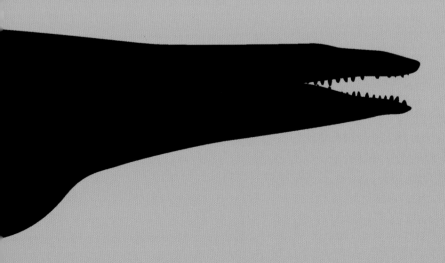

130 Million Years Ago
Present-Day England

The plesiosaurs, born in the Triassic period, were tough and vibrant. Gradually, they toppled the dominance of the Ichthyosauria order and became the new hegemon in the ocean in the Cretaceous period.

The *Leptocleidus* was working hard to practice hunting skills in a large estuary, where many fish were swimming downstream. Still, the hunter felt a little nervous. It opened its mouth to reveal its sharp teeth, with the four flippers moving rhythmically, getting ready for the upcoming hunt.

The one-and-a-half-meter-long *Leptocleidus* was small. To run away from the large plesiosaurs and pliosaurs, they left the vast ocean and lived in the large estuaries, where saltwater and freshwater met. The place was not spacious, and the scenery was a little bland, but the *Leptocleidus* did not mind. After all, it was more important to live to see the next day.

120 Million Years Ago
Present-Day Australia

The living environment was complex and changing. One species may have a certain advantage in one situation, but that could turn into a disadvantage in another. If the species could not understand this, it would soon face mortal danger.

A *Woolungasaurus* was hunted by the ferocious *Kronosaurus*.

The *Woolungasaurus* was trying to catch some small fish and crustaceans to fill its stomach, not expecting to face a *Kronosaurus*. The *Kronosaurus* had a short neck, a large mouth, sharp teeth, explosive strength, and speed. It was a relative of the *Woolungasaurus*—both belonged to the Plesiosauria order—but when it was looking for food, it did not matter whether the hunted was a relative.

Soon, the *Kronosaurus*'s sharp teeth easily bit into the *Woolungasaurus*'s slender neck and small head. The hunted had such tiny structures that the *Kronosaurus* found it easier to catch.

When the *Woolungasaurus* hunted, its slender neck and small head were helpful; they allowed it to hide and attack from a distance. The *Woolungasaurus* did not expect that those features would cause its death.

120 Million Years Ago Present-Day Germany

Contrary to the plesiosaurs' flourishing, the ichthyosaurs were in serious decline. Only a handful of members survived and tried to regain the family's dominance, to no avail.

One hundred twenty million years ago in present-day Germany, the weather was excellent, and the sea was calm. A group of sea turtles had laid eggs and were returning to the ocean. Having new family members was a cheerful prospect. But they did not expect that a dark undercurrent was gathering on their way back home.

Under the surface of the ocean, a *Platypterygius* was patrolling with its family members. Upon seeing the excited sea turtle group, it wasted no time in sending a signal to attack. The group quickly changed into attacking formation, surrounded the turtles. The leader opened its big mouth and bit into the neck of a turtle, and another *Platypterygius* bit the victim's flipper.

The *Platypterygius* was about seven meters long, medium size by the standard of all ichthyosaurs. Their bodies were perfectly shaped to allow for quick movement, and they lived in the ocean for about thirty-five million years, acting as the final guardians of the ichthyosaurs.

110 Million Years Ago
Present-Day Canada

It was late at night, but the *Maiaspondylus* had trouble with sleep.

The once-noisy ocean was quiet at night, with no one playing around, making noise, or hunting. Everything seemed to be wrapped in endless loneliness, intensified by soundless water.

The *Maiaspondylus* twisted its body, hoping to make the two little ones in its belly a little more comfortable. The two were probably expecting something good about the outside world as they were always sending hopeful signals. But the mother was feeling differently. It lived each day in endless despair.

The *Maiaspondylus* looked at its companions who were sleeping nearby. Their bodies were beautiful, elegant like an ichthyosaur should be. They tried to live each day to the fullest as if that day would be the last day of life.

Well, was the last day already here? The *Maiaspondylus* felt sad. The glory of the family had long passed. Former companions were rarely seen in the ocean, and it had to live with a few friends, secluded from everyone else.

Fortunately, it had two children.

Yes, two children. The *Maiaspondylus* felt embarrassed. The unborn children were weak but still full of hope. Why should it give up? No, it had to grit its teeth and carry on, and start from this evening. It should get a good night's sleep for a better tomorrow.

Like the *Platypterygius,* the *Maiaspondylus* was also the last hope of the ichthyosaurs. Unfortunately, these last members of the family did not recover glory. They witnessed the demise of the family.

After the ichthyosaurs disappeared, the ocean was without hegemony for some time. Seizing this opportunity, the mosasaurs stepped forward to become the new hegemon.

110 Million Years Ago
Present-Day Niger

Compared to the *Maiaspondylus*, the *Sarcosuchus* lived a much simpler life. It belonged to the crocodiles, which were flourishing in the peak of the Cretaceous period; even the dominant plesiosaurs avoided them. It was the golden age for the *Sarcosuchus*, and they had a great time in the water.

One hundred ten million years ago, in present-day Niger, a *Sarcosuchus* wandered around a lake.

Not far away, a small pliosaur was hunting a school of small fish. The *Sarcosuchus* watched for a while, then left in boredom. To it, the battle was nothing.

It is no wonder that the eleven-meter-long *Sarcosuchus* thought that way. This giant had scary-looking scales on its back, the longest of which was one meter high. It also had more than one hundred huge and sharp teeth, with amazing bite force. Armed with these weapons, it could easily defeat a dinosaur of the same size. No wonder it had no interest in those poor little fish!

The *Sarcosuchus* wandered elsewhere for a while, and when it came back, the pliosaur had not finished the fight. It sighed and decided to give the pliosaur a hunting lesson. It put its body completely in the water, leaving only two eyes slightly above, and then waited patiently.

Soon, something moved. A *Suchomimus* seemed to be getting near. The *Sarcosuchus* was getting tense for the fight. While the *Suchomimus* lowered its body to drink water, the *Sarcosuchus* shot out of the water like a rocket, opening its crocodile-like mouth and biting into the hind legs of the *Suchomimus*. Its front teeth were extremely sharp, capable of quickly tearing the victim into pieces.

As the *Suchomimus* shrieked in panic, a real battle began with a scream of panic. Not far from the scene, the pliosaur stopped to look back. The school of little fish, who suffered some casualties, caught this break and quickly escaped.

100 Million Years Ago
Present-Day England

They had worked hard, but they could not change the sad ending. The *Acamptonectes* watched the end of everything and did not regret its work. Thanks to everyone in the Ichthyosauria order, the *Acamptonectes* could put up a wonderful finale.

The *Acamptonectes* was trying to swim to the surface of the ocean. It was not chasing a delicious little fish but was trying to look at the stars.

It heard that the sky was like a grand party every night, with the stars putting up shiny glimmers. But it has never seen that scene.

The *Acamptonectes* had always been busy. It was busy preying, fighting for the glory of the family. It had no time to take a good look at the world. Now, it gradually understood that the end was coming, as everything must come to an end. If it could not escape fate, it should be brave and face the end!

The *Acamptonectes* decided not to preoccupy its life with things to do. It wanted to have a talk with the world, with the stars, moon, wind, rain, and everything that it was not paying attention to. It could be a nice way to end the family. At least, the end would be calm and beautiful, leaving no regrets.

The *Acamptonectes* was the last member of the Ichthyosauria order. They lived until ninety-four million years ago and then left the world gracefully. The ichthyosaurs, which lived in the world for one hundred fifty million years, left with them. They brought many wonderful stories to the world, and one should not be saddened by their absence. After a short interlude, new lives would take the stage and continue the ocean's legends.

98 Million Years Ago
Present-Day Italy

The afternoon weather was so hot that it sucked the vitality out of everything. With no wind, the leaves, sand, and stones were all lying motionlessly. The *Aigialosaurus* felt that it was wrapped in heat as if it was in a steamer. It slowly moved towards a river and eventually soaked its slender, snake-like body in the clear water. The water was sluggish, flowing only slowly. The *Aigialosaurus* had to swing the flat tail from time to time to get some current to wash its body.

For the *Aigialosaurus*, living in the water meant a way to get some comfort against the sweltering weather. Its limbs were not suited to swimming. Naturally, it did not expect that its descendants would become hegemons in the water. The simple-minded *Aigialosaurus* did not want to think about the distant future. It preferred to live a happy life as it was.

The *Aigialosaurus* was a peculiar amphibious lizard, generally considered to be the ancestor of the mosasaurs.

93 Million Years Ago
Present-Day Angola

About five million years after the *Aigialo-saurus* entered the water, the mosasaurs finally appeared. At first, they looked like accidental intruders, because their snake-like bodies were a rare sight in the ocean. But that unusual look gave them a majestic aura.

Angolasaurus spent a long time getting here. It stopped at a place where the light was bright. The long travel made its four tiny flippers tired, such that a good night's sleep would be helpful. Well, the residents did not want that to happen too quickly. The *Angolasaurus* did not look like a fish, and its neck was short. The locals had never seen anything like this, and those around were gossiping, making sounds that the *Angolasaurus* did not understand. The *Angolasaurus* thought that it perhaps should say hello first. It twisted its body and danced, picking its usual method of making friends, fully expecting others to join the dance. But that did not happen. Instead, the animals were laughing at it.

Angolasaurus felt both awkward and sad, not sure if it should stop. It looked around in embarrassment, but they gave it looks as if they were looking at a clown. After they had their laugh, they scattered, leaving the snake-like *Angolasaurus* alone. The laughing audience did not know that the *Angolasaurus* was not a clown at all, but it would soon become the most powerful predator in this ocean.

93 Million Years Ago
Present-Day America

In fact, the *Angolasaurus* was not the only "clown." Around the same time, another strange-looking species was living in the ocean of present-day North America.

If the *Russellosaurus* knew the story of *Angolasaurus*, it would probably give it a long lecture, "You tell them something like this: 'Don't meddle with us tough guys.' Look at our slender and flexible bodies, and our short and powerful limbs. These showed how we crawled on land. Those guys had never seen land and had no idea

what kind of magical world it is. Let them look at our long and sharp teeth. If they annoy us, we would bite them and not let go, with the sharp teeth getting deep into their bodies. Well, don't interrupt, I haven't finished . . ."

Although the *Russellosaurus* was nagging and often had a bad temper, it was speaking the truth. The new *Mosasauridae* family was a force to acknowledge.

93 Million Years Ago
Present-Day Morocco

Let us go to Africa, where we could find members of the *Mosasauridae* family who lived there in the Early Cretaceous period. Soon after they appeared, the emerging mosasaurs began to spread globally.

Having laid eggs on land, a *Tethysaurus* twisted its lizard-like body to return to the water. It crawled slowly, not because laying eggs was especially tiring, but it was a slow mover on land. Apart from laying eggs and rest, it spent most of the time in the water, so its ability to walk on land gradually faded.

The *Tethysaurus* had spent a long time returning to the water. As soon as it entered the water, it quickly drew its limbs close to its body and swam as flexibly as a sea snake. The free and happy swimmer once again returned to its preferred habitat.

92 Million Years Ago
Present-Day America

One choice could often completely change destiny. The mosasaurs changed theirs by taking the first step towards the ocean.

The land was shrinking, and food was scarce. The *Dallasaurus* found it increasingly difficult to survive. There were only two ways: succumb to reality, continue to live a difficult life, or start on a completely new road—into the ocean, where the competition was less intense.

The *Dallasaurus* had been thinking about this problem for a long time. As it became more and more difficult to fill its belly, it finally chose the challenging road. A successful adventure needs courage and luck, and fortunately, it had both. Like *Angolasaurus* and *Tethysaurus,* it adapted to living in water, contributing to the *Mosasauridae* family's dominance in the ocean.

Looking at their sizes, the earliest members of the *Mosasauridae* family were not particularly outstanding. But they played their cards right to use nature's gifts. When they were born, the polar ice melted, creating many new inland seas in present-day Central America and northwest Africa. The mosasaurs seized the opportunity to enter the vast and empty sea. Food was abundant, and the competition was mild, so the mosasaurs developed rapidly and saw their first peak.

92 Million Years Ago
Present-Day America

The *Brachauchenius* and the *Dallasaurus* lived in the same ocean. The *Brachauchenius* always thought that the slim and peculiar *Dallasaurus* was a joke. Compared to its twelve-meter-long body, the *Dallasaurus* almost seemed an unworthy dwarf. The *Brachauchenius* felt that only something like itself was worthy of living in the ocean. It did not know that it was at the end of the Pliosauroidea suborder; the pliosaurs in this region of present-day North America would soon die out. The *Mosasauridae* family, including the *Dallasaurus*, would begin to develop.

The *Brachauchenius* whizzed through the ocean. It opened its huge and terrifying mouth, brandishing its numerous sharp teeth, trying to devour all animals that got in its way, thinking that it would continue to reign powerfully. Unfortunately, its arrogance would eventually lead to its demise.

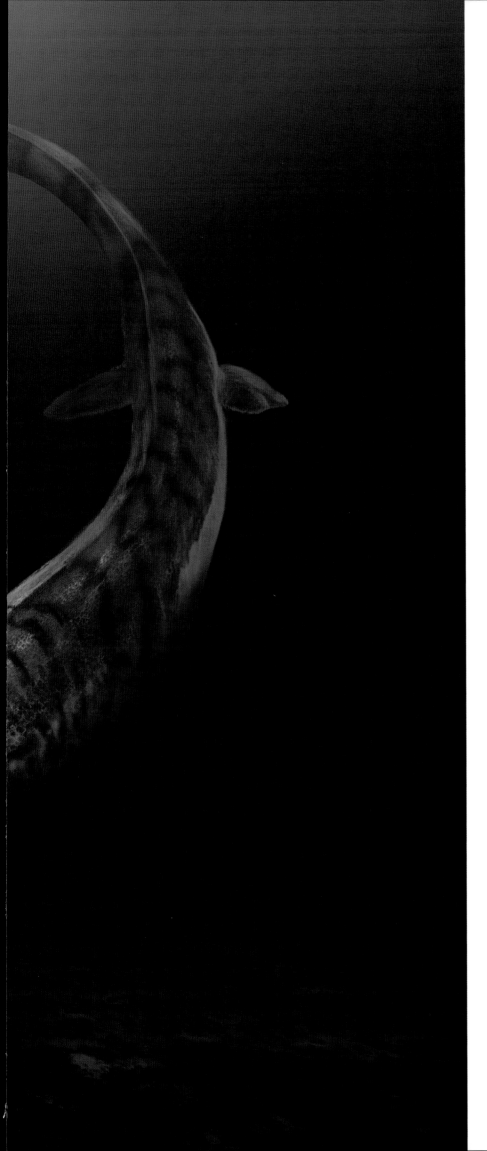

85 Million Years Ago
Present-Day America

The *Clidastes* was daring. Since it was only about three meters long, living in the shallow sea where the competition was gentle was more sensible. But it insisted on going to the dangerous deep sea. Its limbs were too small to move its body powerfully, and they were only good enough to steer; its body was also tiny, with the small ribs holding up a small belly, which could not carry much food. But it was not worried. It had a head shaped like a bullet, which could guide it to move fast in the ocean. Its vision was excellent, so it always found prey and enemies in time and used its spear-like jaw and sharp teeth to tear them to pieces. Of course, the most important thing is that it had the courage and insight that many other guys did not have. That was how it lived bravely in the dangerous deep sea.

85 Million Years Ago
Present-Day Mexico

Compared to the *Clidastes*, the *Amphekepubis* was bigger but lazy. It wished that the prey will be delivered to its doorstep without it moving a finger. So, it reasoned that a good way to achieve that was to live in the shallow sea, where competitors were few. That would solve everything, or would it?

Eighty-five million years ago in present-day Mexico, an *Amphekepubis* and its companions swam leisurely in a shallow sea. Being nine meters long, they were the giants. They had what they wanted: if they stayed in one place and opened their mouths, from time to time, a prey running in panic would deliver itself to them.

These *Amphekepubis* had been living like this for a long time, and their bodies became obese due to lack of exercise. But they were not worried and enjoyed what they could get without putting in any effort.

However, the good times did not last long. The *Amphekepubis* were not the only ones who enjoyed a comfortable life. More and bigger guys heard about their lives and rushed to this shallow sea, making the place overcrowded and the competition tougher. Often, even its nine-meter-long body did not stop more powerful opponents from attacking the *Amphekepubis.* Unfortunately, its fat body was defenseless against the enemies.

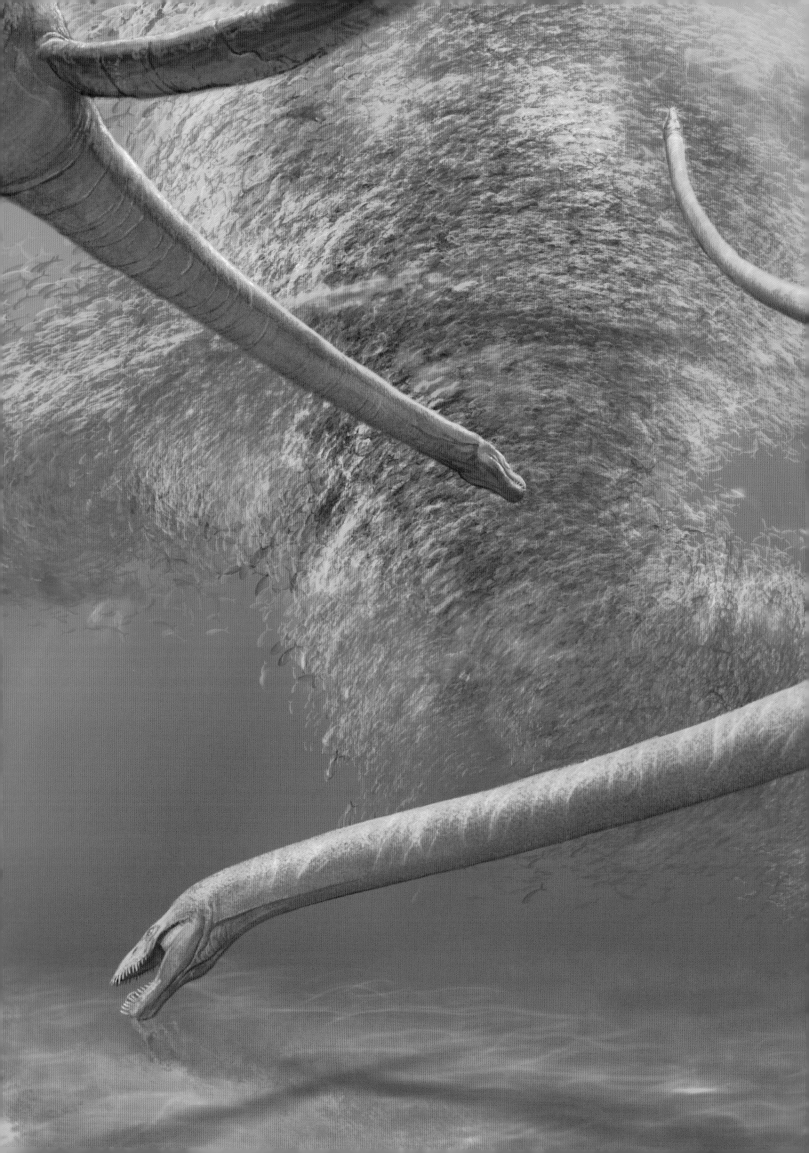

85 Million Years Ago
Present-Day America

In hunting battles, numbers are not always everything. Strategy, skill, and the "tools" used were the most crucial factors in determining the outcome. These reasons seemed to account for the victories won by the plesiosaurs.

The *Elasmosaurus* led its family members to charge at a school of fish in the ocean. Their long necks were shooting like arrows. The disturbed fish quickly formed a whirlpool in the water, trying to obscure the enemy's vision, but it did not seem to work.

Although the *Elasmosaurus* did not have overwhelming numbers, their skillful attack formation and their unique body structure had already decided the outcome of this battle.

85 Million Years Ago
Present-Day America

One does not always play with the leading role in life's drama. The plesiosaurs won the contest against the ichthyosaurs, but they did not have time to enjoy the taste of victory before meeting the mosasaurs, who seemed to come from nowhere. The plesiosaur could not afford to be arrogant but had to concentrate and fight. It was a tough battle, and one careless mistake could doom the family to submission.

The *Styxosaurus* did not expect its life to be difficult. It was eleven meters long, not small for a plesiosaur, but it had become a weak, hunted animal in this ocean. The ocean was the same, but ambitious intruders had arrived.

The *Styxosaurus* tried to fight, but it had been scarred many times, showing that death was just around the corner. It had to give up, try to be patient, and live a low-key life. This made its life less dangerous, and it was happy.

But fate was determined to make its life difficult. That very same morning, it met a powerful member of the mosasaurs, a *Tylosaurus*. It was a bloodthirsty intruder, one that made the *Styxosaurus* shudder.

Without hesitation, the *Tylosaurus* opened its giant mouth and went straight at the *Styxosaurus*. *Styxosaurus* hesitated but quickly dived at the fastest speed. Fortunately, it swam past the *Tylosaurus*!

The *Styxosaurus* escaped this disaster and swam forward as quickly as it could. It was not sure whether the *Tylosaurus* would give up and even less certain whether it would be as lucky the next time.

85 Million Years Ago
Present-Day America

Some members of the *Mosasauridae* family were huge, but their initial claim to fame was not their large size. Prestige always requires something special, and the *Ectenosaurus* got this prestige because of its head.

A bright light beam ran through the clear sea and shone on the *Ectenosaurus*. The pattern on its back glowed softly, blending in with the seafloor. It raised its head proudly, getting it illuminated by another ray.

The defining feature of the *Ectenosaurus* was its long and pointed head, especially its narrow and long snout, which was different from all the members of the mosasaurs. Residents would always hide quickly if they saw that head moving. Now, the sunshine made that head more conspicuous.

However, the *Ectenosaurus* never worried about its head foiling its hunting plans. The skull had many holes, which reduced the weight of the head. Its body was agile, making its movement in the water fast. So, even those guys who tried to hide at first sight of the predator seldom made it to safety.

85 Million Years Ago
Present-Day Canada

Desire destroys everything: a calm life, a happy mood, and any good stuff that you can think of. Fortunately, the *Kourisodon* chose to live a simple life, even though it was from a powerful and famous family.

The day was about to begin, and the *Kourisodon* stretched and smiled with charm. This day started with a simple breakfast, a few small fish plus a few crustaceans. The *Kourisodon* was a member of the mighty mosasaurs. It was four meters long, not a giant, and some would think that it could hunt targets more worthy than small fish. The *Kourisodon,* however, felt that anything that filled its stomach was good enough. By being less picky about food, it could spend more time enjoying other things, like going on a long trip or observing the life of a coral. Others might think it was mad, but this *Kourisodon* enjoyed life this way, thinking that this was the only way to peace and happiness.

85 Million Years Ago
Present-Day America

With the decline of the Pliosauroidea sub-order, the once-powerful Plesiosauroidea suborder faced its imminent demise. They did not give up hope but bravely made changes. One group of them abandoned the characteristic long neck and the slender body, trying to move even more quickly in the water. They were the *Polycotylus* genus of the Plesiosauroidea suborder, and they lived for about three million years. During this short period, they made their footprints all over the world.

Eighty-five million years ago in present-day America, a *Polycotylus* was chasing an ammonite in the ocean. Although the prey was small, the *Polycotylus* was dedicated to the pursuit. It was pregnant but still hardworking. This was its way of life, taking everything seriously.

Unlike most members of the Plesio-sauroidea suborder, the *Polycotylus* had a particularly wide body, with a short neck and a large head. It was a fast hunter, and its large flippers gave it a powerful boost.

83 Million Years Ago
Present-Day America

Most of the original members of the *Mosasauridae* family were concentrated in the shallow seas of present-day North America. The family expanded, but the earliest of them continued to live like their ancestors. Until about eighty-three-and-a-half million years ago, the mosasaurs entered in the second period of rapid development. They moved from shallow sea to deep sea, spreading to present-day Europe, Africa, Oceania, and Antarctica. Many large species appeared, and it was the heyday of the family.

The prosperity of the family brought not only glory but also troubles. For example, *Hainosaurus* felt that many huge and fierce guys suddenly appeared, making the pressure of its survival significantly greater.

In order to survive in the fierce competition, *Hainosaurus* had to improve itself, including getting in shape, improving swimming skills, and trying to be less picky about food. The last point was important. In such a complicated environment, having a large appetite was fundamental for survival.

83 Million Years Ago
Present-Day Russia

The sunlight shone upon the ocean, turning it into a magnificent palace, like the ones in eighteenth-century Europe. A *Dollosaurus* was swimming in this palace. It looked a little nervous, looking back every few steps.

The *Dollosaurus* was certainly not looking for enemies or prey. It was about ten meters long, a strong fighter, and one that never got nervous about a fight or a hunt.

What was it worried about? Well, there was a female *Dollosaurus* behind it, someone who was beautiful, noble, and elegant, like a princess. The male's heart began pumping wildly.

The *Dollosaurus* was a member of the *Mosasauridae* family. The only thing in the ocean that made this one uneasy was love!

83 Million Years Ago
Present-Day America

The silliest thing in the world was to make yourself a replicate of others. *Selmasaurus* understood this point too well. Otherwise, its life would be terribly dark!

Selmasaurus had been in this shallow and narrow sea since it was born. Of course, it wanted to go somewhere far, to visit the broader and deeper ocean, and to kiss the unruly waves. But it could not. Unlike other members of the mosasaurs, it did not have a joint between its upper and lower jaws that could make the mouth open larger. This handicap made it impossible to hunt large prey, while its relatives could. It would probably starve to death in the deep ocean. Therefore, it had to stay in the shallow sea, feeding on small fish and crustaceans.

However, *Selmasaurus* never felt sad or inferior because of this. It thought of this not as a flaw, but the defining feature of *Selmasaurus*. Because of this small mouth, it could quietly live here, where the competition was mild, and fill its stomach easily. Not having to worry about livelihood, it could play around the beautiful plants. Its leisure was something that other members of the mosasaurs would never get.

80 Million Years Ago
Present-Day America

After the last glimmer of the setting sun, the ocean turned blue again. The *Plioplatecarpus* had been watching the sunset until the gorgeous colors left the ocean, and it returned to calm darkness. Then, it slowly swayed its tail, diving into the blue ocean.

Unlike other mosasaurs, who took a lot of time to fill their stomachs, the smart *Plioplatecarpus* always got enough food with little effort, so it had time to enjoy viewing the sunset.

The *Plioplatecarpus* was a latecomer in the family but soon rose to prominence. This was entirely due to their wisdom. The *Plioplatecarpus* had a large brain and was the most intelligent member of the mosasaurs. They defeated many enemies with their ingenuity and became the ocean's hegemon.

75 Million Years Ago
Present-Day Netherlands

The development of the *Mosasauridae* family was not always smooth. The lower temperature caused glaciers to appear on the poles, and the ocean level fell. Some shallow seas became land, destroying the habitats for some mosasaurs.

But they did not become extinct. On the contrary, about seventy-five million years ago was another peak in their development. Numerous new members appeared and tried to make their rule of the ocean permanent.

The *Carinodens*, who lived during the third peak of the *Mosasauridae* family, did not look overbearing at all. It was petite, only three meters long, probably one of the smallest members of the family. Although it could not deal with large prey, it was good at catching small sea urchins and clams.

The small *Carinodens* served its purpose. A truly prosperous family must occupy every section of the food chain. Therefore, even the smaller members are indispensable.

75 Million Years Ago
Present-Day America

One could learn from others' experiences but not repeat someone else's success. The *Globidens* knew this too well. To survive, it did not choose to use its body size or structure but specialized in choosing its food. Adult *Globidens* no longer ate fast-moving fish but picked the slow-moving animals like crustaceans, ammonites, or lobsters. Eating these alternative foods made competition less fierce. They did pay the price; by hunting only the slower prey, they would never become good swimmers. Their reward was that the family flourished.

Two chubby *Globidens* were eagerly chasing two small ammonites. Their spherical teeth were like small mortars, perfect for crushing the prey's hard shell.

75 Million Years Ago
Present-Day England

The *Liodon* also had special teeth. Unlike the *Globidens*'s spherical teeth, theirs were known for being replaceable.

The world of *Liodon* had no dentists. When they broke or lost their teeth while fighting or hunting, new ones grew automatically to fill the gaps, in a way like present-day sharks.

Two ten-meter-long *Liodon* were fighting over prey. Both were giants, and neither wanted to give up. Each one had its teeth biting the other's neck. The gushing blood soon dyed the oceanwater crimson, but both *Liodon* were not ending the fight. Instead, they tried to bury the sharp teeth deeper. They were not worried that such a big force would break the teeth. After all, new ones would grow out soon.

72 Million Years Ago
Present-Day America

In the Late Cretaceous period, the plesiosaur was no longer as prosperous as it once was, but it did not die out. One species grew to twenty-five meters long, and they cruised in the ocean like giant submarines, flaunting their strength.

It was still early. The sun had not risen, and the ocean was still surrounded by a dreamy grayish hue, while the *Cimoliasaurus* had already set off.

Its body was too big. As it swam, every stroke would make a huge sound. The sleepy animals, who would have been easy targets, escaped because of the sound.

That was only a trivial inconvenience for the *Cimoliasaurus* because it knew that the ocean would become livelier during the day when the prey could no longer hide in darkness and it could eat all it wanted. Now, it should pick a nice spot to hide and wait patiently while everyone sleeps.

70 Million Years Ago
Present-Day Nigeria

There are many ways to hunt. Most commonly, hunters use either their eyesight or smell to search for prey. The more unusual hunters wait and ambush or use feigning attacks against unsuspecting targets. Of course, a good method is one that works, and a clever hunter is one with catches.

The hunter we are going to meet is the smart *Goronyosaurus*. Its eyes were small, and its vision was mediocre. So, it did not use its vision to sense prey but used its touch.

Touch? Yes, exactly. In a muddy environment, its highly sensitive snout often was successful.

A *Goronyosaurus* was hiding in the dark, waiting for prey to approach. It closed its eyes and let its body be guided by its extraordinary perception. Suddenly, it felt prey was swimming towards it. It immediately looked up and opened its mouth, waiting for the prey to deliver itself. Even if the prey was a member of the same *Mosasauridae* family, it would not let it go.

70 Million Years Ago
Present-Day America

When the ocean level declined and turned inland seas to land, it was a disaster, but the looming darkness soon went away. Members of the *Mosasauridae* family soon found a better place to live, the more spacious deep sea, where food was abundant. Everyone was hopeful for a better life in the new habitat.

In addition to them, the plesiosaurs felt the same way. The few surviving members, such as the *Elasmosaurus,* were active once again. They were large and aggressive, capable of fighting on equal footing against the fierce mosasaurs.

No one was thinking about imminent doom, and they all lived freely.

A ten-meter-long *Plesiotylosaurus* was confidently waiting for an opponent. It did not matter who the challenger was, whether it was a plesiosaur, such as the *Aphrosaurus*, *Fresnosaurus*, or *Hydrotherosaurus*, or a mosasaur, like the *Plotosaurus*. It was looking forward to beating any new challenger so the world would know that it was the ruler of this sea.

70 Million Years Ago
Present-Day America

Well, the *Plotosaurus* would not be afraid of the *Plesiotylosaurus*. It was thirteen meters long, but it did not brag about that. It preferred to let everyone know that it was a perfect swimmer. Its body was as streamlined as a big fish, the perfect shape for living in the ocean; it had a crescent tail, resembling a fish's, which provided it with incredible power; its scales were like the shark's shield scale, overcoming the sea water's resistance; it had sharp vision to see everything in the dark ocean. Were these features not enough to defeat the *Plesiotylosaurus*?

172

The *Plotosaurus* swam arrogantly through the ocean, like a gigantic and majestic fish. The *Plotosaurus* represented almost the perfect evolution of terrestrial creatures' adaptation in the ocean. If it were not for the extinction disaster, it might evolve into something indistinguishable from fish, swimming in present-day oceans.

However, that hypothetical situation did not happen. The disaster fell without warning, at a time when the mosasaurs, plesiosaurs, and other marine creatures had high hopes.

66 Million Years Ago
Present-Day Netherlands

The other creatures could brag for as long as they wanted, but the ultimate glory of the mosasaurs and even of the entire ocean belonged to this one.

It was the most powerful animal in the ocean, one that quickly conquered the ocean with unparalleled power and wisdom. This was a miracle unmatched to this day.

It was the ultimate member of the *Mosasauridae* family, the *Mosasaurus*.

The fifteen-meter-long *Mosasaurus* attacked a two-meter-long *Archelon*. The strong teeth of the *Mosasaurus* easily smashed the primitive carapace of the latter, stabbing into its organs, turning the water instantly turbid. In the upper jaw of the *Mosasaurus*'s mouth, another row of teeth was tearing the victim's skin and flesh. Before the *Archelon* felt the pain, it was torn into pieces.

50 Million Years Ago
Present-Day Pakistan

The arrival of the *Mosasauridae* family was the peak of the marine reptiles.

It would be difficult for us to remember the days that they were little lizards. Now, looking at the ocean's hegemon, the change seemed to have been completed in an instant. The older inhabitants had no time to think about what happened before a group of large monsters suddenly ruled their lives. The mosasaurs did not disappoint the ocean, which created one legend after another.

However, such glory was too short. Before the *Mosasauridae* family had time to savor it, a terrible disaster fell upon them. Life did not bestow special favor on the legendary *Mosasauridae* family; it destroyed them when they were strongest and most powerful, and it was a dreadful fate which they could not escape.

A mass extinction from sixty-six million years ago destroyed everything. The mosasaurs, along with most other marine reptiles, disappeared from the ocean. Their departure was regrettable. The consolation was that they had always been the best swimmers until the day they left earth.

The ocean was not as lonely as it was five hundred million years ago. Soon, a new group of animals followed the step of the predecessors, moving from land to ocean. They were whales, the first mammals to appear in the ocean.

An *Ambulocetus* was looking curiously at the ocean from nearby.

It was less than three meters long and spent most of the time walking on the shore. The *Abmulocetus* did not yet know that in twenty million years it would lead the cetaceans to evolve and to become stronger and more powerful than the mosasaurs.

35 Million Years Ago
Present-Day America

Under the warm sunshine, the ocean was always welcoming. It was getting ready for a quiet moment. For billions of years, countless animals had come and left. After the fierce mosasaurs disappeared, the ocean was expecting solitude. Unexpectedly, a group of whales, equally active and ambitious, came. They quickly dissociated themselves from land and lived in the ocean comfortably. Look, the chubby *Zygorhiza* was moving its fins to enjoy a leisurely life. The ocean would once again be full of life.

New residents would bring new ways. There would be bloody
fights about life and death, and there would be sunny days and
relaxed times. Everyone would have time to think about what kind
of life they should live and what kind of world they should create.

Countless worlds have risen and declined.

Now, the ocean was expecting the new master, the whales, to
create more legends.

Chronicle of Events in the Age of Ancient Sea Monsters

The Chaotic Triassic Period

Two hundred fifty million years ago, a global extinction destroyed most invertebrates in the ocean and more than 70 percent of reptiles on land. The vast ocean was temporarily silent. The surviving reptiles seized this opportunity, preparing themselves to return to the ocean for more living space and more abundant food.

Two hundred fifty million years ago, *Hupehsuchus* appeared in the ocean. It was one of the earliest aquatic reptiles.

Two hundred forty-five million years ago, the petite *Chaohusaurus* unexpectedly moved to the unpredictable sea. No one could foresee that this would be the first step of the ocean's profound changes to follow. The glorious Ichthyosauria order was born.

Two hundred forty-four million years ago, the ocean welcomed the *Thalattoarchon*. The emergence of this giant proved that the ichthyosaurs had developed rapidly in a short period of time, and the ocean lives were recovering. More species were appearing for the first time.

Two hundred forty-two million years ago, the Ichthyosauria order, born in the early Triassic period, saw their first explosive growth in the Middle Triassic period, with the *Phantomosaurus* appearing at this time.

Two hundred forty million years ago, the birth of the *Qianosuchus* represented the emergence of the Archosauria order, a large group of marine and land animals.

Two hundred forty million years ago, the *Nothosaurus* joined the ocean. Like modern seals, it was one of the founders of the Nothosauroidea order, whose feet had evolved into paddles to move them forward in the water.

Two hundred forty million years ago, *Anshunsaurus* was born, which belonged to the Thalattosauria order. They looked like lizards and had flat tails that moved them forward.

Two hundred forty million years ago, with its wide body and carapace, the *Glyphoderma* was a new genus of aquatic reptiles. They represented the Placodontia order, an unusual group. Part of their bodies looked a tortoise, and other parts looked like a lizard.

Two hundred thirty-two million years ago, some members of the Protorosauria order, which was born in the Permian period, chose to live in the Triassic ocean. Among them was the *Tanystropheus*. Their extremely long necks were unique in the ocean at the time. They were large, so the thriving ichthyosaurs usually did not feed on them.

Two hundred thirty million years ago, the *Pachypleurosaurus*, which was closely related to the nothosaurs, began to join its relatives in the ocean. The newcomers did not find adapting to ocean life easy, so they became low-key and practiced living skills in the shallow sea near the shore.

Two hundred twenty-five million years ago, the ichthyosaurs had made great progress in their territorial expansion, an attempt which started a long time ago. The *Shastasaurus*, which appeared at this time, spread its members to the world's oceans.

Two hundred twenty million years ago, the emergence of *Californosaurus* brought unprecedented impact to the ichthyosaurs. Unlike the earlier ichthyosaurs' stout, lizard-like body shapes, the *Californosaurus* were almost perfectly streamlined, like fish. This change would play a crucial role in the survival of the ichthyosaurs.

Two hundred ten million years ago, after experiencing explosive growth and reaching a peak in their geographical spread, aquatic reptiles suddenly experienced another decline in the Late Triassic period. The Ichthyosauria order and the Plesiosauria order (the latter evolved from the nothosaurs) made it through the Triassic period and continued to thrive in the Cretaceous period. Apart from them, the rest of the aquatic reptiles quickly declined.

The Silent Jurassic Period

One hundred ninety-eight million years ago, the dolphin-like *Temnodontosaurus* was showing off its perfectly streamlined body to other animals in the ocean. The Ichthyosauria order now looked completely different from the original, primitive form. This was an important turning point, and afterward, they became fully adapted to marine life.

One hundred eighty-three million years ago, the Plesiosauria order, which evolved from the nothosaurs, began to rise. They shared the vast sea alongside the ichthyosaurs.

One hundred seventy million years ago, unlike most plesiosaurs living in the ocean, the *Yuzhoupliosaurus* chose freshwater lakes. A new living environment had both opportunities and challenges.

One hundred sixty-seven million years ago, a group of highly efficient predators, the marine crocodiles, wished to grab its share of the ocean from the ruling ichthyosaurs and the plesiosaurs. The marine crocodiles used to be terrestrial, but as the ocean eroded land, they had to look for new habitat. Losing their familiar habitat gave them a strong desire to survive.

One hundred sixty million years ago, the family of the plesiosaurs, which appeared later than the Ichthyosauria order, developed strongly. They not only rapidly spread the family into the ocean of the world, but also grew rapidly. Almost overnight, large and fierce monsters appeared.

One hundred fifty million years ago, facing a strong opponent in the Pliosauroidea suborder of the Plesiosauria order, the Ichthyosauria order slowly declined. Although several of them have expanded their location to present-day South America, by the end of the Jurassic, only a few families remained active in the ocean.

One hundred forty-five million years ago, when the ichthyosaurs and the plesiosaurs were fighting for dominance, the marine crocodiles seized the opportunity to grow up quietly. They completely adapted to the life in the ocean in a short time and became globally distributed, using their tricks to become the world's top predators.

The Volatile Cretaceous Period

One hundred thirty million years ago, after entering the Cretaceous period, the few remaining members of the Ichthyosauria order were unable to rule the waves of the ocean. The Plesiosauria order, born in the Trissaic period, had more vitality. The powerful plesiosaurs were confidently replacing the ichthyosaurs as the hegemon of the ocean.

One hundred twenty million years ago, as one of the last members of the family, the unfortunate *Platypterygius* witnessed the demise of the ichthyosaurs.

One hundred ten million years ago, marine crocodiles climbed to the peak of their lives. They were fierce; even the mighty plesiosaurs chose not to fight them.

Ninety-three million years ago, *Angolasaurus* appeared as the first member of the *Mosasauridae* family in the ocean. Its arrival foretold that the ocean's pattern will face major adjustments.

Ninety-two million years ago, the newly-born mosasaurs showed amazing power. In a short period of time, they drastically changed their body structures and rapidly expanded their habitat. In an environment of less competition and plenty of food, they quickly had the first peak of explosive growth, laying a solid foundation to become the ocean's top rulers.

Eighty-five million years ago, the plesiosaurs won the battle against the ancient ichthyosaurs but suffered a Waterloo defeat against the mosasaurs.

Eighty-three-and-a-half million years ago, the *Mosasauridae* family saw their second rapid development. They moved from shallow sea to deep sea. From present-day North America to Europe, Africa, Oceania, Antarctica, and other places, many large mosasaurs appeared, ushering in the heyday of the family.

Seventy-five million years ago, a recent decline in sea level had reduced their habitat, but the emergence of many new members recovered the *Mosasauridae* family's morale, and the family saw their third peak of development.

Seventy million years ago, the plesiosaurs experienced a short-lived revival. The *Elasmosaurus* and a few others became active.

Sixty-six million years ago, one of the most powerful animals in the ocean, the *Mosasaurus*, was born. It quickly conquered the ocean with unparalleled power and wisdom, leaving unmatched legends.

Sixty-six million years ago, Earth once again saw an extinction event. The splendid mosasaurs and the reviving plesiosaurs died out in the disaster, but their enduring legends continued to inspire future animals.

References

1. Piñeiro, G., A. Ramos, C. Goso, F. Scarabino, and M. Laurin. 2012. "Unusual environmental conditions preserve a Permian mesosaur-bearing Konservat-Lagerstätte from Uruguay." *Acta Palaeontologica Polonica* 57 (2): 299–318.

2. Canoville, Aurore, and Michel Laurin. 2010. "Evolution of humeral micro-anatomy and lifestyle in amniotes, and some comments on paleobiological inferences." *Biological Journal of the Linnean Society* 100: 384–406.

3. Modesto, S. P. 2010. "The postcranial skeleton of the aquatic parareptile *Mesosaurus tenuidens* from the Gondwanan Permian." *Journal of Vertebrate Paleontology* 30 (5): 1378–95.

4. Chen, Xiaohong, P. Martin Sander, Long Cheng, and Xiaofeng Wang. 2013. "A New Triassic Primitive Ichthyosaur from Yuanan, South China." *Acta Geologica Sinica* (English Edition) 87 (3): 672–77.

5. Young, C. C., and Z. Dong. 1972. "On the Triassic aquatic reptiles of China." *Memoires of the Nanjing Institute of Geology and Paleontology* 9: 1–34.

6. Liezhu, Chen. 1985. "Ichthyosaurs from the lower Triassic of Chao County." *Anhui Regional Geology of China* 15: 139–46.

7. Mazin, J.-M., V. Suteethorn, E. Buffetaut, J.-J. Jaeger, and R. Helmckeingavat. 1991. "Preliminary description of *Thaisaurus chonglakmanii* n. g., n. sp., a new ichthyopterygian (Reptilia) from the Early Triassic of Thailand." *Comptes Rendus de l'Académie des Sciences* Serie II, 313: 1207–12.

8. Motani, R., and H. You. 1998. "The forefin of *Chensaurus chaoxianensis* (Ichthyosauria) shows delayed mesopodial ossification." *Journal of Paleontology* 72 (1): 133–36.

9. Motani, R., and H. You. 1998. "Taxonomy and limb ontogeny of *Chaohusaurus geishanensis* (Ichthyosauria), with a note on the allometric equation." *Journal of Vertebrate Paleontology* 18 (3): 533–40.

10. Maisch, M. W. 2001. "Observations on Triassic ichthyosaurs. Part VII. New data on the osteology of *Chaohusaurus geishanensis* YOUNG & DONG, 1972 from the Lower Triassic of Anhui (China)." *Neues Jahrbuch für Geologie und Paläontologie, Abhandlungen* 219 (3): 305–17.

11. Motani, Ryosuke, Da-yong Jiang, Andrea Tintori, Olivier Rieppel, and Guan-bao Chen. 2014. "Terrestrial Origin of Viviparity in Mesozoic Marine Reptiles Indicated by Early Triassic Embryonic Fossils." *PLoS ONE* 9 (2): e88640.

12. Maisch, M. W. 2010. "Phylogeny, systematics, and origin of the Ichthyosauria – the state of the art." *Palaeodiversity* 3: 151–214.

13. Motani, R. 1999. "Phylogeny of the Ichthyopterygia." *Journal of Vertebrate Paleontology* 19 (3): 473–96.

14. Motani, R., N. Minoura, and T. Ando. 1998. "Ichthyosaurian relationships illuminated by new primitive skeletons from Japan." *Nature* 393 (6682): 255–57.

15. Shikama, T., T. Kamei, and M. Murata. 1977. "Early Triassic Ichthyosaurus, *Utatsusaurus hataii* Gen. et Sp. Nov., from the Kitakami Massif, Northeast Japan." *Science Reports of the Tohoku University Second Series (Geology)* 48 (1–2): 77–97.

16. Motani, R. 1997. "New information on the forefin of *Utatsusaurus* hataii (Ichthyosauria)." *Journal of Paleontology* 71 (3): 475–79.

17. Cuthbertson, R. S., A. P. Russell, and J. S. Anderson. 2013. "Reinterpretation of the cranial morphology of *Utatsusaurus hataii* (Ichthyopterygia) (Osawa Formation, Lower Triassic, Miyagi, Japan) and its systematic implications." *Journal of Vertebrate Paleontology* 33 (4): 817–30.

18. Motani, R. 1996. "Redescription of the dental features of an early Triassic ichthyosaur, *Utatsusaurus hataii*." *Journal of Vertebrate Paleontology* 17 (1): 39–44.

19. Chen, X. H., R. Motani, L. Cheng, D. Y. Jiang, and O. Rieppel. 2014. "The Enigmatic Marine Reptile *Nanchangosaurus* from the Lower Triassic of Hubei, China and the Phylogenetic Affinities of Hupehsuchia." *PLoS ONE* 9 (7): e102361.

20. Fröbischa, N. B., J. R. Fröbischa, P. M. Sanderb, L. Schmitzc, and O. Rieppel. 2013. "Macropredatory ichthyosaur from the Middle Triassic and the origin of modern trophic networks." *Proceedings of the National Academy of Sciences* 110 (4): 1393–97.

21. Nesbitt, S. J. 2011. "The early evolution of archosaurs: relationships and the origin of major clades." *Bulletin of the American Museum of Natural History* 352: 1–292.

22. Diedrich, C. 2009. "The vertebrates of the Anisian/Ladinian boundary (Middle Triassic) from Bissendorf (NW Germany) and their contribution to the anatomy, palaeoecology, and palaeobiogeography of the Germanic Basin reptiles." *Palaeogeography, Palaeoclimatology, Palaeoecology* 273 (1): 1–16.

23. Rieppel, O. 1994. "The status of the sauropterygian reptile *Nothosaurus juvenilis* from the Middle Triassic of Germany." *Palaeontology* 37 (4): 733–45.

24. Li, J., and O. Rieppel. 2004. "A new nothosaur from Middle Triassic of Guizhou, China." *Vertebrata PalAsiatica* 42 (1): 1–12.

25. Jiang, W., M. W. Maisch, W. Hao, Y. Sun, and Z. Sun. 2006. "*Nothosaurus yangjuanensis* n. sp. (Reptilia, Sauropterygia, *Nothosauridae*) from the middle Anisian (Middle Triassic) of Guizhou, southwestern China." *Neues Jahrbuch für Geologie und Paläontologie, Monatshefte* 2006 (5): 257–76.

26. Shang, Q.-H. 2006. "A new species of *Nothosaurus* from the early Middle Triassic of Guizhou, China." *Vertebrata PalAsiatica* 44 (3): 237–49.

27. Albers, P. C. H. 2005. "A new specimen of *Nothosaurus marchicus* with features that relate the taxon to *Nothosaurus winterswijkensis*." *Vertebrate Paleontology* 3 (1).

28. Klein, N., and P. C. H. Albers. 2009. "A new species of the sauropsid reptile *Nothosaurus* from the Lower Muschelkalk of the western Germanic Basin, Winterswijk, The Netherlands." *Acta Palaeontologica Polonica* 54 (4): 589–98.

29. Schroder, H. 1914. "Wirbeltiere der Riidersdorfer Trias." *Abhandlungen der Preussischen Geologischen Landesanstalt, Neue Folge* 65: 1–98.

30. Rieppel, O., and R. Wild. 1996. "A revision of the genus *Nothosaurus* (Reptilia: Sauropterygia) from the Germanic Triassic with comments on the status of *Conchiosaurus clavatus*." *Fieldiana* New Series 34.

31. Liu, J., and O. Rieppel. 2005. "Restudy of *Anshunsaurus huangguoshuensis* (Reptilia: Thalattosauria) from the Middle Triassic of Guizhou, China." *American Museum Noviates* 3488.

32. Rieppel, O., J. Liu, and C. Li. 2006. "A new species of the thalattosaur genus *Anshunsaurus* (Reptilia: Thalattosauria) from the Middle Triassic of Guizhou Province, southwestern China." *Vertebrata PalAsiatica* 44 (4): 285–96.

33. Cheng, L., X. Chen, and C. Wang. 2007. "A new species of Late Triassic *Anshunsaurus* (Reptilia: Thalattosauria) from Guizhou Province." *Acta Geologica Sinica* 81 (10): 1345–51.

34. Zhao, L.-J., C. Li, J. Liu, and T. He. 2008. "A new armored placodont from the Middle Triassic of Yunnan Province, Southwestern China." *Vertebrata PalAsiatica* 46 (3): 171–77.

35. Palmer, D., ed. 1999. *The Marshall Illustrated Encyclopedia of Dinosaurs and Prehistoric Animals*. London: Marshall Editions.

36. Ketchum, Hilary F., and Roger B. J. Benson. 2011. "A new pliosaurid (Sauropterygia, Plesiosauria) from the Oxford Clay Formation (Middle Jurassic, Callovian) of England: evidence for a gracile, longirostrine grade of Early-Middle Jurassic pliosaurids." *Special Papers in Palaeontology* 86: 109–29.

37. Sepkoski, Jack. 2002. "A compendium of fossil marine animal genera (entry on Reptilia)." *Bulletins of American Paleontology* 364: 560.

38. Renesto, S. 2005. "A new specimen of *Tanystropheus* (Reptilia Protorosauria) from the Middle Triassic of Switzerland and the ecology of the genus." *Rivista Italiana di Paleontologia e Stratigrafia* 111 (3): 377–94.

39. Tschanz, K. 1988. "Allometry and Heterochrony in the Growth of the Neck of Triassic Prolacertiform Reptiles." *Paleontology* 31 (4): 997–1011.

40. Motani, R., H. You, and C. McGowan. 1996. "Eel like swimming in the earliest ichthyosaurs." *Nature* 382: 347–48.

41. Jiang, D.-Y., L. Schmitz, W.-C. Hao, and Y.-L. Sun. 2006. "A new mixosaurid ichthyosaur from the Middle Triassic." *Journal of Vertebrate Paleontology* 85 (1): 32–36.

42. Schmitz, L. 2010. "The taxonomic status of *Mixosaurus nordenskioeldii*." *Journal of Vertebrate Palaeontology* 25: 983–85.

43. Motani, R. 1999. "Phylogeny of the Ichthyopterygia." *Journal of Vertebrate Paleontology* 19: 473–96.

44. Maisch, Michael W., and Andreas T. Matzke. 2000. "The Ichthyosauria." *Stuttgarter Beiträge zur Naturkunde: Serie B* 298: 1–159.

45. Da-Yong Jiang, Lars Schmitz, Wei-Cheng Hao, and Yuan-Lin Sun (2006). "A new mixosaurid Ichthyosaur from the Middle Triassic of China." *Journal of Vertebrate Paleontology* 26 (1): 60–69.

46. Jiang, D.-Y., O. Rieppel, N. C. Fraser, R. Motani, W.-C. Haoa, A. Tintorie, Y.-L. Suna, and Z.-Y. Suna. 2011. "New information on the protorosaurian reptile *Macrocnemus fuyuanensis* Li et al., 2007, from the Middle/Upper Triassic of Yunnan, China." *Journal of Vertebrate Paleontology* 31 (6): 1230–37.

47. Cheng, Yen-Nien, Tamaki Sato, Xiao-Chun Wu, and Chun Li. 2006. "First complete pistosauroid from the Triassic of China." *Journal of Vertebrate Paleontology* 26 (2): 501–4.

48. Shang, Qing-Hua, and Chun Li. 2009. "On the occurrence of the ichthyosaur Shastasaurus in the Guanling Biota (Late Triassic), Guizhou, China." *Vertebrata PalAsiatica* 47 (3): 178–93.

49. Nicholls, E. L., and M. Manabe. 2004. "Giant ichthyosaurs of the Triassic - a new species of *Shonisaurus* from the Pardonet Formation (Norian: Late Triassic) of British Columbia." *Journal of Vertebrate Paleontology* 24 (4): 838–49.

50. Sander, P. M., X. Chen, L. Cheng, and X. Wang. 2011. "Short-Snouted Toothless Ichthyosaur from China Suggests Late Triassic Diversification of Suction Feeding Ichthyosaurs," edited by Leon Classens. *PLoS ONE* 6 (5): e19480.

51. Motani, R., T. Tomita, E. Maxwell, D. Jiang, and P. Sander. 2013. "Absence of Suction Feeding Ichthyosaurs and Its Implications for Triassic Mesopelagic Paleoecology." *PLoS ONE* 8 (12): e66075.

52. Ji, C., D.-Y. Jiang, R. Motani, W.-C. Hao, Z.-Y. Sun, and T. Cai. 2013. "A new juvenile specimen of *Guanlingsaurus* (Ichthyosauria, Shastasauridae) from the Upper Triassic of southwestern China." *Journal of Vertebrate Paleontology* 33 (2): 340–48.

53. Maisch, M. W., and A. T. Matzke. 2006. "The braincase of *Phantomosaurus neubigi* (Sander, 1997), an unusual ichthyosaur from the Middle Triassic of Germany." *Journal of Vertebrate Paleontology* 26 (3): 598–607.

54. Xiaofeng, W., G. H. Bachmann, H. Hagdorn, P. M. Sander, G. Cuny, C. Xiaohong, W. Chuanshang, C. Lide, C. Long, M. Fansong, and X. U. Guanghong. 2008. "The Late Triassic Black Shales of the Guanling Area, Guizhou Province, South-West China: A Unique Marine Reptile and Pelagic Crinoid Fossil Lagerstätte." *Palaeontology* 51 (1): 27–61.

55. Rieppel, O. C. 2000. "Paraplacodus and the phylogeny of the Placodontia (Reptilia: Sauropterygia)." *Zoological Journal of the Linnean Society* 130 (4): 635–39.

56. Rieppel, O. C., and R. T. Zanon. 1997. "The interrelationships of Placodontia. " *Historical Biology* 12 (3–4): 211–27.

57. Merriam, J. C. 1902. "Triassic Ichthyopterygia from California and Nevada." *Bulletin of the Department of Geology of the University of California* 3 (4).

58. Motani R. 2000. "Rulers of the Jurassic seas." *Scientific American* 283 (6): 52–59.

59. Martin, J. E., V. Fischer, P. Vincent, and G. Suan. 2010. "A longirostrine *Temnodontosaurus* (Ichthyosauria) with comments on Early Jurassic ichthyosaur niche partitioning and disparity." *Palaeontology* 55 (5): 995–1005.

60. McGowan, C. 1996. "Giant ichthyosaurs of the Early Jurassic." *Canadian Journal of Earth Sciences* 33 (7): 1011–21.

61. McGowan, C. 1995. "Temnodontosaurus risor is a Juvenile of *T. platyodon* (Reptilia: Ichthyosauria)." *Journal of Vertebrate Paleontology* 14 (4): 472–79.

62. Sander, P. M. 2000. "Ichthyosauria: their diversity, distribution, and phylogeny." *Paläontologische Zeitschrift* 74 (1): 1–35.

63. Buchholtz, Emily A. 2000. "Swimming styles in Jurassic Ichthyosaurs." *Journal of Vertebrate Paleontology* 21 (1): 61–73.

64. Motani, R. 2005. "Evolution of fish-shaped reptiles (Reptilia: Ichthyopterygia) in their physical environments and constraints." *Annual Review of Earth and Planetary Sciences* 33: 395–420.

65. Scheyer, Torsten M., Carlo Romano, Jim Jenks, and Hugo Bucher. 2014. "Early Triassic Marine Biotic Recovery: The Predators' Perspective." *PLoS ONE* 9 (3): e88987.

66. McGowan, C. 1974. "A revision of the longipinnate ichthyosaurs of the Lower Jurassic of England, with descriptions of two new species (Reptilia, Ichthyosauria)." *Life Sciences Contributions*, Royal Ontario Museum 97.

67. Maisch, Michael W., and Andreas T. Matzke. 2003. "Observations on Triassic ichthyosaurs. Part XII. A new Lower Triassic ichthyosaur genus from Spitzbergen." *Neues Jahrbuch für Geologie und Paläontologie Abhandlungen* 229 (3): 317–38.

68. Benton, M. J., and M. A. Taylor. 1984. "Marine reptiles from the Upper Lias (Lower Toarcian, Lower Jurassic) of the Yorkshirecoast." *Proceedings of the Yorkshire Geological Society* 44: 399–429.

69. Martill, D. M. 1993. "Soupy Substrates: A Medium for the Exceptional Preservation of Ichthyosaurs of the Posidonia Shale (Lower Jurassic) of Germany." *Kaupia - Darmstädter Beiträge zur Naturgeschichte* 2: 77–97.

70. Caine, Hannah, and Michael J. Benton. 2011. "Ichthyosauria from the Upper Lias of Strawberry Bank, England." *Palaeontology* 54 (5): 1069–93.

71. Maxwell, E. E., M. S. Fernández, and R. R. Schoch. 2012. "First Diagnostic Marine Reptile Remains from the Aalenian (Middle Jurassic): A New Ichthyosaur from Southwestern Germany," edited by Andrew A. Farke. *PLoS ONE* 7 (8): e41692.

72. Fischer, V., E. Masure, M. S. Arkhangelsky, and P. Godefroit. 2011. "A new Barremian (Early Cretaceous) ichthyosaur from western Russia." *Journal of Vertebrate Paleontology* 31 (5): 1010–25.

73. Drunkenmiller, Patrick S., and Erin E. Maxwell. 2010. "A new Lower Cretaceous (lower Albian) ichthyosaur genus from the Clearwater Formation, Alberta, Canada." *Canadian Journal of Earth Sciences* 47 (8): 1037–53.

74. Smith, Adam S. 2007. "Anatomy and systematics of the Rhomaleosauridae (Sauropterygia, Plesiosauria)." Ph.D. thesis, University College Dublin.

75. Smith, Adam S., and Gareth J. Dyke. 2008. "The skull of the giant predatory pliosaur *Rhomaleosaurus cramptoni:* implications for plesiosaur phylogenetics." *Naturwissenschaften* 95 (10): 975–80.

76. Benson, Roger B. J., Hilary F. Ketchum, Leslie F. Noè, and Marcela Gómez-Pérez. 2011. "New information on *Hauffiosaurus* (Reptilia, Plesiosauria) based on a new species from the Alum Shale Member (Lower Toarcian: Lower Jurassic) of Yorkshire, UK." *Palaeontology* 54 (3): 547–71.

77. Ketchum, Hilary F., and Roger B. J. Benson. 2011. "A new pliosaurid (Sauropterygia, Plesiosauria) from the Oxford Clay Formation (Middle Jurassic, Callovian) of England: evidence for a gracile, longirostrine grade of Early-Middle Jurassic pliosaurids." *Special Papers in Palaeontology* 86: 109–29.

78. Smith, Adam S., and Peggy Vincent. 2010. "A new genus of pliosaur (Reptilia: Sauropterygia) from the Lower Jurassic of Holzmaden, Germany." *Palaeontology* 53 (5): 1049–63.

79. Fischer, V., M. Guiomar, and P. Godefroit. 2011. "New data on the palaeobiogeography of Early Jurassic marine reptiles: the Toarcian ichthyosaur fauna of the Vocontian Basin (SE France)." *Neues Jahrbuch für Geologie und Paläontologie, Abhandlungen* 261 (1): 111–27.

80. Reisdorf, A. G., M. W. Maisch, and A. Wetzel. 2011. "First record of the leptonectid ichthyosaur *Eurhinosaurus longirostris* from the Early Jurassic of Switzerland and its stratigraphic framework." *Swiss Journal of Geosciences* 104 (2): 211–24.

81. Zhang, Y. 1985. "A new plesiosaur from Middle Jurassic of Sichuan Basin." *Vertebrata PalAsiatica* 23 (3): 65–89.

82. Fernández, Marta S. 1994. "A new long-snouted ichthyosaur from the Early Bajocian of Neuquén Basin, Argentina." *Ameghiniana* 31 (3): 283–90.

83. Fernández, Marta S. 1999. "A new ichthyosaur from the Los Molles Formation (Early Bajocian), Neuquén Basin, Argentina." *Journal of Paleontology* 73 (4): 677–81.

84. Young, M. T. 2007. "The evolution and interrelationships of *Metriorhynchidae* (Crocodyliformes, Thalattosuchia)." *Journal of Vertebrate Paleontology* 27 (3): 170A.

85. Gasparini, Z., D. Pol, and L. A. Spalletti. 2006. "An unusual marine crocodyliform from the Jurassic-Cretaceous boundary of Patagonia." *Science* 311: 70–73.

86. Wilkinson, L. E., M. T. Young, and M. J. Benton. 2008. "A new metriorhynchid crocodilian (Mesoeucrocodylia: Thalattosuchia) from the Kimmeridgian (Upper Jurassic) of Wiltshire, UK." *Palaeontology* 51: 1307–33.

184

87. Cau, Andrea, and Federico Fanti. 2010. "The oldest known metriorhynchid crocodylian from the Middle Jurassic of North-eastern Italy: *Neptunidraco ammoniticus* gen. et sp. nov." *Gondwana Research* 19 (2): 550–56.

88. Gandola, R., E. Buffetaut, N. Monaghan, and G. Dyke. 2006. "Salt glands in the fossil crocodile *Metriorhynchus*." *Journal of Vertebrate Paleontology* 26 (4): 1009–10.

89. Fernández, M., and Z. Gasparini. 2008. "Salt glands in the Jurassic metriorhynchid *Geosaurus*: implications for the evolution of osmoregulation in Mesozoic crocodyliforms." *Naturwissenschaften* 95 (1): 79–84.

90. Forrest, R. 2003. "Evidence for scavenging by the marine crocodile *Metriorhynchus* on the carcass of a plesiosaur." *Proceedings of the Geologists' Association* 114: 363–66.

91. Young, Mark T., and Marco Brandalise de Andrade. 2009. "What is *Geosaurus*? Redescription of *Geosaurus giganteus* (Thalattosuchia: *Metriorhynchidae*) from the Upper Jurassic of Bayern, Germany." *Zoological Journal of the Linnean Society* 157 (3): 551–85.

92. Fernández, M. 2007. "Redescription and phylogenetic position of *Caypullisaurus* (Ichthyosauria: *Ophthalmosauridae*)." *Journal of Paleontology* 81 (2): 368–75.

93. Fischer, V., A. Clement, M. Guiomar, and P. Godefroit. 2011. "The first definite record of a Valanginian ichthyosaur and its implications on the evolution of post-Liassic Ichthyosauria." *Cretaceous Research* 32 (2): 155–63.

94. Maxwell, E. E. 2010. "Generic reassignment of an ichthyosaur from the Queen Elizabeth Islands, Northwest Territories, Canada." *Journal of Vertebrate Paleontology* 30 (2): 403–15.

95. Fischer, Valentin, Michael W. Maisch, Darren Naish, Ralf Kosma, Jeff Liston, Ulrich Joger, Fritz J. Krüger, Judith Pardo Pérez, Jessica Tainsh, and Robert M. Appleby. 2012. "New Ophthalmosaurid Ichthyosaurs from the European Lower Cretaceous Demonstrate Extensive Ichthyosaur Survival across the Jurassic-Cretaceous Boundary." *PLoS ONE* 7 (1): e29234.

96. Schumacher, B. A., K. Carpenter, and M. J. Everhart. 2013. "A new Cretaceous Pliosaurid (Reptilia, Plesiosauria) from the Carlile Shale (middle Turonian) of Russell County, Kansas." *Journal of Vertebrate Paleontology* 33 (3): 613–28.

97. Benson, R. B. J., and P. S. Druckenmiller. 2013. "Faunal turnover of marine tetrapods during the Jurassic-Cretaceous transition." *Biological Reviews* 89 (1): 1–23.

98. Benson, R. B. J., M. Evans, A. S. Smith, J. Sassoon, S. Moore-Faye, Hilary F. Ketchum, and Richard Forrest. 2013. "A Giant Pliosaurid Skull from the Late Jurassic of England." *PLoS ONE* 8 (5): e65989.

99. Wilhelm, B. C., and F. R. O'Keefe. 2010. "A New Partial Skeleton of a Cryptocleidoid Plesiosaur from the Upper Jurassic Sundance Formation of Wyoming." *Journal of Vertebrate Paleontology* 30 (6): 1736–42.

100. Zammit, M., C. B. Daniels, and B. P. Kear. 2008. "Elasmosaur (Reptilia: Sauropterygia) neck flexibility: Implications for feeding strategies." *Comparative Biochemistry and Physiology Part A: Molecular & Integrative Physiology* 150 (2): 124–30.

101. Martill, D. M., M. A. Taylor, and K. L. Duff. 1994. "The trophic structure of the biota of the Peterborough Member, Oxford Clay Formation (Jurassic), UK." *Journal of the Geological Society* 151 (1): 173–94.

102. Cicimurri, D. J., and M. J. Everhart. 2001. "An Elasmosaur with Stomach Contents and Gastroliths from the Pierre Shale (Late Cretaceous) of Kansas." *Transactions of the Kansas Academy of Science* 104 (3–4): 129–43.

103. Bardet, N., and M. Fernández. 2000. "A new ichthyosaur from the Upper Jurassic lithographic limestones of Bavaria." *Journal of Paleontology* 74 (3): 503–11.

104. Massare, J. A. 1988. "Swimming capabilities of Mesozoic marine reptiles; implications for method of predation." *Paleobiology* 14 (2): 187–205.

105. Massare, J. A. 1987. "Tooth morphology and prey preference of Mesozoic marine reptiles." *Journal of Vertebrate Paleontology* 7 (2): 121–37.

106. Smith, A. S., and G. J. Dyke. 2008. "The skull of the giant predatory pliosaur *Rhomaleosaurus cramptoni*: implications for plesiosaur phylogenetics." *Naturwissenschaften* 95 (10): 975–80.

107. Sachs, S. 2004. "Redescription of *Woolungasaurus glendowerensis* (Plesiosauria: *Elasmosauridae*) from the Lower Cretaceous of Northeast Queensland." *Memoirs of the Queensland Museum* 49 (2): 713–31.

108. Maxwell, Erin E., and Michael W. Caldwell. 2006. "A new genus of ichthyosaur from the Lower Cretaceous of western Canada." *Palaeontology* 49 (5): 1043–52.

109. Maxwell, Erin E., and Michael W. Caldwell. 2003. "First record of live birth in Cretaceous ichthyosaurs: closing an 80 million year gap." *Proceedings of the Royal Society of London B* 270: 104–7.

110. Le Loeuff, J., E. Métais, D. B. Dutheil, J. L. Rubino, E. Buffetaut, F. Lafont, L. Cavin, F. Moreau, H. Tong, C. Blanpied, and A. Sbeta. 2010. "An Early Cretaceous vertebrate assemblage from the Cabao Formation of NW Libya." *Geological Magazine* 147 (5): 750–59.

111. Head, J. J. 2001. "Systematics and body size of the gigantic, enigmatic crocodyloid *Rhamphosuchus crassidens*, and the faunal history of Siwalik Group (Miocene) crocodylians." *Journal of Vertebrate Paleontology* 21 (Supplement to No. 3): 59A.

112. Erickson, G. M., and C. A. Brochu. 1999. "How the 'terror crocodile' grew so big." *Nature* 398: 205–6.

113. Sereno, Paul. C., and Stephen L. Brusatte. 2008. "Basal abelisaurid and carcharodontosaurid theropods from the Lower Cretaceous Elrhaz Formation of Niger." *Acta Paleontologica Polonica* 53 (1): 15–46.

114. Fischer, Valentin, Michael W. Maisch, Darren Naish, Ralf Kosma, Jeff Liston, Ulrich Joger, Fritz J. Krüger, Judith Pardo Pérez, Jessica Tainsh, and Robert M. Appleby. 2012. "New Ophthalmosaurid Ichthyosaurs from the European Lower Cretaceous Demonstrate Extensive Ichthyosaur Survival across the Jurassic-Cretaceous Boundary." *PLoS ONE* 7 (1): e29234.

115. Dutchak, Alex R., and Michael W. Caldwell. 2009. "A redescription of *Aigialosaurus* (=*Opetiosaurus*) *bucchichi* (Kornhuber, 1901) (Squamata: *Aigialosauridae*) with comments on mosasauroid systematics." *Journal of Vertebrate Paleontology* 29 (2): 437–52.

116. Everhart, M. J. 2005. *Oceans of Kansas: A Natural History of the Western Interior Sea*. Indiana University Press.

117. Tanimoto, M. 2005. "Mosasaur remains from the Upper Cretaceous Izumi Group of southwest Japan." *Geologie en Mijnbouw (Netherlands Journal of Geosciences)* 84 (3): 373–78.

118. O'Keefe, F.R., and L. M. Chiappe. 2011. "Viviparity and K-selected life history in a Mesozoic marine plesiosaur (Reptilia, Sauropterygia)." *Science* 333 (6044): 870–73.

119. Albright, L. B., III, D. D. Gillette, and A. L. Titus. 2007. "Plesiosaurs from the Upper Cretaceous (Cenomanian-Turonian) Tropic Shale of southern Utah, Part 2: Polycotylidae." *Journal of Vertebrate Paleontology* 27 (1): 41–58.

120. Lindgren, Johan. 2005. "The first record of *Hainosaurus* (Reptilia: Mosasauridae) from Sweden." *Journal of Paleontology* 79 (6): 1157–65.

121. Christiansen, P., and N. Bonde. 2002. "A new species of gigantic mosasaur from the Late Cretaceous of Israel." *Journal of Vertebrate Paleontology* 22 (3): 629–44.

122. Lindgren, J., H. F. Kaddumi, and M. J. Polcyn. 2013. "Soft tissue preservation in a fossil marine lizard with a bilobed tail fin." *Nature Communications* 4: 2423.

123. LeBlanc, Aaron R. H., Michael W. Caldwell, and Nathalie Bardet. 2012. "A new mosasaurine from the Maastrichtian (Upper Cretaceous) phosphates of Morocco and its implications for mosasaurine systematics." *Journal of Vertebrate Paleontology* 32 (1): 82–104.

124. Russel, Dale, 1975. "A new species of *Globidens* from South Dakota, and a review of globidentine mosasaurs." *Fieldiana Geology* 33 (13).

125. Bardet, N., X. Pereda Suberbiola, M. Iarochene, M. Amalik, and B. Bouya. 2005. "Durophagous Mosasauridae (Squamata) from the Upper Cretaceous phosphates of Morocco, with description of a new species of *Globidens*." *Geologie en Mijnbouw (Netherlands Journal of Geosciences)* 84 (3): 167–75.

126. O'Keefe, F. Robin, and Hallie P. Street. 2009. "Osteology of the Cryptoclidoid Plesiosaur *Tatenectes laramiensis*, with Comments on the Taxonomic Status of the *Cimoliasauridae*." *Journal of Vertebrate Paleontology* 29 (1): 48–57.

127. Mulder, E. W. A. 1999. "Transatlantic latest Cretaceous mosasaurs (Reptilia, Lacertilia) from the Maastrichtian type area and New Jersey." *Geologie en Mijnbouw (Netherlands Journal of Geosciences)* 84 (3): 315–20.

128. Mulder, E. W. A. 2004. "Maastricht Cretaceous finds and Dutch pioneers in vertebrate palaeontology." In *Dutch Pioneers of the Earth Sciences*, edited by J. L. R. Touret and R. P. W. Visser. Royal Netherlands Academy of Arts and Sciences, Amsterdam 165–76.

Index

186

ZHAO Chuang and YANG Yang
&
PNSO's Scientific Art Projects Plan: Stories on Earth (2010–2070)

ZHAO Chuang and YANG Yang are two professionals who work together to create scientific art. Mr. ZHAO Chuang, a scientific artist, and Ms. YANG Yang, an author of scientific children's books, started working together when they jointly founded PNSO, an organization devoted to the research and creation of scientific art in Beijing on June 1, 2010. A few months later, they launched Scientific Art Projects Plan: Stories on Earth (2010–2070). The plan uses scientific art to create a captivating, historically accurate narrative. These narratives are based on the latest scientific research, focusing on the complex relationships between species, natural environments, communities, and cultures. The narratives consider the perspectives of human civilizations while exploring Earth's past, present, and future. The PNSO founders plan to spend sixty years to do research and create unique and engaging scientific art and literature for people around the world. They hope to share scientific knowledge through publications, exhibitions, and courses. PNSO's overarching goal is to serve research institutions and the general public, especially young people.

PNSO has independently completed or participated in numerous creative and research projects. The organization's work has been shared with and loved by thousands of people around the world. PNSO collaborates with professional scientists and has been invited to many key laboratories around the world to create scientific works of art. Many works produced by PNSO staff members have been published in leading journals, including *Nature*, *Science*, and *Cell*. The organization has always been committed to supporting state-of-the-art scientific explorations. In addition, a large number of illustrations completed by PNSO staff members have been published and cited in hundreds of well-known media outlets, including the *New York Times*, the *Washington Post*, the *Guardian*, *Asahi Shimbun*, the *People's Daily*, BBC, CNN, Fox News, and CCTV. The works created by PNSO staff members have been used to help the public better understand the latest scientific discoveries and developments. In the public education sector, PNSO has held joint exhibitions with scientific organization, including the American Museum of Natural History and the Chinese Academy of Sciences. PNSO has also completed international cooperation projects with the World Young Earth Scientist Congress and the Earth Science Matters Foundation, thus helping young people in different parts of the world understand and appreciate scientific art.

KEY PROJECTS

I. Darwin: An Art Project of Life Sciences
*The models are all life-sized and are based on fossils found around the world
1.1 Dinosaur fossils
1.2 Pterosaur fossils
1.3 Aquatic reptile fossils
1.4 Ancient mammals of the Cenozoic Era
1.5 Chengjiang biota: animals in the Early Cambrian from fossils discovered in Chengjiang, Yunnan, China
1.6 Jehol biota: animals in the Mesozoic Era from fossils discovered in Jehol, Western Liaoning, China
1.7 Early and extinct humans
1.8 Ancient animals that coexisted with early and extinct humans
1.9 Modern humans
1.10 Animals of the *Felidae* family
1.11 Animals of the *Canidae* family
1.12 Animals of the Proboscidea order

Books in this series

Age of Dinosaurs

Age of Pterosaurs

Age of Ancient Sea Monsters

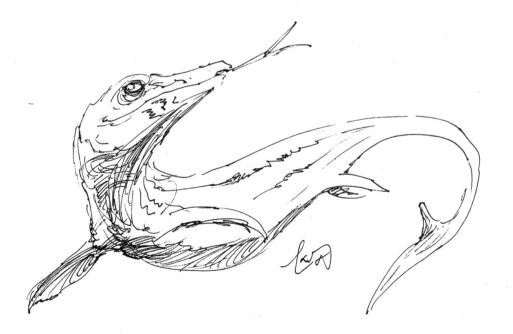